FINGERPRINTING SOFTWARE DEFINED NETWORKS AND CONTROLLERS

I. Introduction

1.1 Background

Cyber warfare is present in the United States' 2014 quadrennial defense review, and the act of countering cyber attacks is paramount for investment [1]. The necessity of network infrastructure parallels the requirement in ensuring a sufficient defense against adversaries willing to exploit any vulnerability. From corporate entities losing sensitive client information, money, and customer trust, to nation states capturing national defense intelligence, cyber efficacy remains prevalent in today's society. The threat of cyber attacks and interest in securing our cyber borders continually increases as is shown in both current events as well as from security analysts [2][3][4][5].

Enhancing cyber security starts with maintaining an inventory of authorized and unauthorized devices considering attackers are constantly scanning for new unprotected systems that join a potentially sensitive network [6]. In addition to new systems, new technologies that join the network are also lucrative for attackers given any lack of documented security testing may keep unknown vulnerabilities from discovery. New technology presents new avenues for zero-day engineering, which allows an adversary unmitigated capabilities during the life of the zero-day [7].

New technology currently growing in the network environment includes the use of Software Defined Networking (SDN). SDN promises to remove the vendor-specific requirements placed on a network administrator by transforming network switches into simple packet forwarding devices, while placing the brains of the network into logically centralized software [8]. From this centralized location, network administrators need

1

not concern themselves with the underlying vendor-specific implementations in order to accomplish high-level abstract goals. Configuration is simplified. Considering the technology is new, security implications for the increased flexibility have not been fully tested. It is important to identify how the network changes with this radical shift in network topology. Determining what the adversary gains from the network administrators' increased convenience allows defensive updates to guard against cyber threats.

1.2   Problem Statement

This research attempts to identify information that is unintentionally offered to a network attacker when SDN is used within a small network, and demonstrates the feasibility of uniquely identifying the software managing the SDN environment. With positive identification of the software controlling the SDN environment (a process known as fingerprinting), an attacker can then search for existing vulnerabilities or attempt to develop custom attacks against the logically centralized software. Preventing an attacker's discovery of the network controller assists in thwarting the attacker's reconnaissance, ultimately inhibiting the attacker's capabilities. If an attacker is unable to uniquely identify a target, then the list of available vulnerabilities at the attacker's disposal is limited, and the attacker's threat is minimized. Focusing on assessing whether fingerprinting is possible, this research attempts to identify first when SDN is deployed, and then proceeds to gather intelligence in the form of unique features that describe the SDN controller software. The end of this collection of features occurs when the SDN software is successfully fingerprinted.

The methods of collecting data are restricted to the methods available to an attacker. An attacker is assumed to have a presence in the network in the form of a connection to the network switching fabric. The attacker can also communicate to other end hosts on the network, including a host that is another point of presence for the attacker (i.e., the attacker can have two points of presence on the network to communicate between).

1.3 Goals and Hypothesis

The objective of this thesis is to develop and evaluate a process for extracting features to uniquely identifying the controller supporting an SDN environment. The first goal of this research is to construct a set of features extensive enough to uniquely identify each known SDN controller. The next goal includes ensuring that each feature is obtainable by a client connected to the SDN environment. It is hypothesized that a process can be created that adds each new SDN controller into a table of SDN controller features, and that this table can be used to identify an unknown SDN controller discovered in an SDN environment by a client connected to that environment.

1.4 Approach

A simulation of a simple SDN environment is created using MiniNet [25]. Features are programmed into the SDN controller software and then observed by a client connected to the SDN environment. The observed features and true values are compared for accuracy. The same process is performed using network equipment to validate the results obtained from the simulated environment.

1.5 Assumptions

The experiments within this research assume a client has access to a port connected to a network environment. Once an SDN software controller is uniquely identified, the action of searching for documented vulnerabilities and exploiting the SDN controller is beyond the scope of this research. Research is limited to information gathering.

1.6 Contributions

This thesis contributes to the body of research in SDN. Specific contributions include a process for distinguishing SDN environments from traditional environments, the set of features observable by a client within an SDN environment, as well as the process for

3

fingerprinting SDN controllers through the collection and categorization of features in a feature table.

1.7 Thesis Overview

This thesis is organized into five chapters. Chapter 2 defines SDN and presents relevant research needed to understand new threats created by the addition of SDN technology. The methodology for evaluating the fingerprinting process for an SDN environment is explained in Chapter 3. The experimental results and analysis of collected data are presented in Chapter 4. Finally, Chapter 5 summarizes this research and concludes with the results of each experiment, along with an explanation of the significance and suggested future work.

II. Literature Review

This chapter defines SDN, and reviews the relevant bodies of research needed to understand and implement an SDN environment. Section 2.1 introduces a high-level concept of SDN. Section 2.2 defines each component that combine to make up a software-defined network. Section 2.3 explains current uses, growth, and potential of future SDN applications. Next, Section 2.4 provides relevant research in Software Defined Networking specific to network security concerns, followed by Section 2.5, which describes current research in security applications designed to mitigate various security concerns. Section 2.6 reviews research conducted on the latency effects of adopting SDN as well as the potential for latency analysis in gathering information about the SDN environment. This chapter concludes with an explanation of how SDN can be constructed within a network to apply these mitigating techniques.

2.1 Software Defined Networking Defined

Within a computer network, switching devices contain three abstractly organized services known as planes [9]. As shown in Figure 2.1, these planes consist of the forwarding plane (also referred to as the data plane), the control plane, and the management plane. Within Figure 2.1, the left portion shows a traditional network where each box represents a switching device that independently contains all three previously described planes. The software-defined network on the right shows three switches (which only require forwarding capabilities) connected to a controller (which provides control/management plane functionalities). It is the communication between these planes that results in the internals of a network switch. The three planes are typically found within a single device. SDN aims at separating the control plane (inclusive to the management plane) from the forwarding plane, and placing this functionality within a separate device known as a controller.

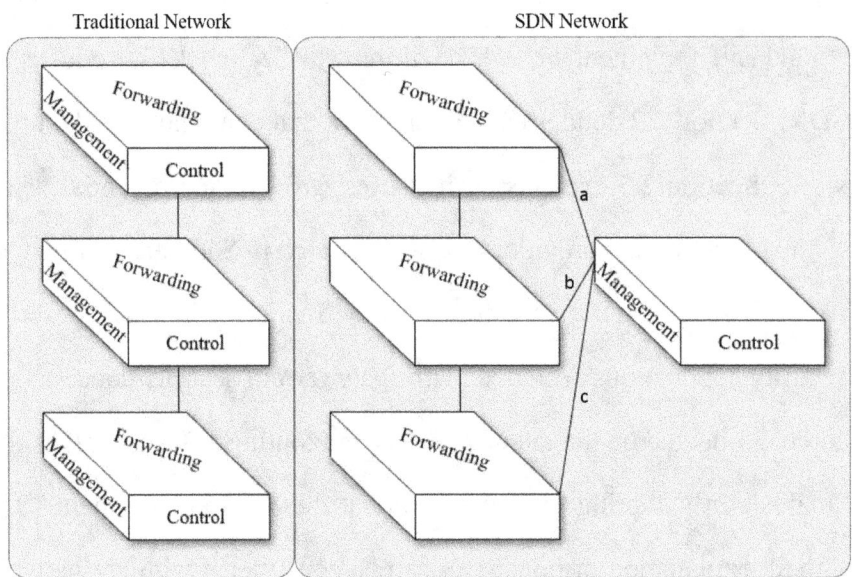

Figure 2.1: SDN Planes

2.1.1   The Forwarding Plane.

The forwarding plane of a switch receives incoming Ethernet frames, and forwards those frames onto a port determined by rules created by the control plane. When a frame is first received on the wire to a specific port of a switch, the forwarding plane calculates various sanity checks for the incoming frame. Such sanity checks include sizing, alignment, and checksum validation [9].   A response to a failed sanity check is implementation dependent, but often results in the frame being dropped. A successful sanity check results in the frame proceeding to the forwarding lookup process. This lookup process is known as the "fast-path" for frames because the lookup process' purpose is specialized in moving the frame to the correct destination quickly, needing only to extract the destination information from the frame [9].

The forwarding lookup process can be implemented in many ways depending on a vendor's proprietary design. The implementation possibilities include performing a lookup within software, hardware-accelerated software (e.g., Graphic Processing Unit (GPU)s or Central Processing Unit (CPU)s), commodity hardware (i.e., a network processor), and specialized hardware (i.e., application-specific integrated circuits).

The forwarding process concludes with a set of specific actions. Actions include forwarding the frame, replicating the frame, dropping the frame, modifying the frame, counting the frame (i.e., keeping a record of frame information for statistical purposes), and queuing the frame. The frame may also be destined for a process running locally on the switch (e.g., for Link Layer Discovery Protocol (LLDP), or Spanning Tree Protocol (STP) updates), in which case the frame will be "punted" from the forwarding lookup process and processed by the route processor [9]. The act of moving a locally-destined frame is accomplished through an internal communication channel.

In the event that a destination port does not match any forwarding rule entry for a particular frame, SDN protocol dictates that the frame is then sent to the control plane for further processing [9][10][11]. The communication between the forwarding plane and the control plane is accomplished through a standardized protocol.

2.1.2   *The Control Plane.*

The control plane is responsible for creating the forwarding table that is utilized by the forwarding plane to send a packet from source (ingress) port to destination (egress) port. This table of entries is populated once a stable topology of the network is established. Methods of informing the control plane about network topology include programming the logic directly or through network protocols (such as link-state or distance-vector algorithms). A packet arriving at the control plane from the forwarding plane is processed, whereby the packet's information may result in a modification of the forwarding plane's forwarding table.

2.1.3   The Management Plane.

The control plane is considered a superset of the management plane, which is responsible for network administration that may result in control plane behavior modification due to administrative policies (such as enforcing access control lists). While the control plane modifies the forwarding table of the forwarding plane, the management plane manages the policies that are enforced by the control plane.

2.1.4   Separation of Planes.

Separating the control plane from the forwarding plane allows control decisions of forwarding behavior to exist on a logically centralized, and thus easily managed, location. Instead of each network switch independently containing all three planes, the switches specialize in performing forwarding-plane functions, leaving the decision of creating rules to a separate controller running specific control-plane software. Having each switch only perform forwarding-plane functions simplifies the processor logic and has the added benefit of commoditizing the switching hardware [10].

Moving the control plane to a separate controller effectively reduces the number of logical control planes within a network to one, rather than equal to the number of switches in the network. By having a single control plane for multiple switches, a network administrator can program flow-control capabilities (such as traffic-shaping) from one logical location, rather than requiring that administrator to touch every switch independently with disparate functionality.

Additionally, other benefits of separating the control and forwarding planes typically result from the innovation allowed by modularity: each plane can be improved independently.   Independent improvements reduce complexity, also increasing stability within a hardware design.   Technological improvements also reduce the price per performance.

The following sections break down software-defined networking into each of its components: the protocol used between devices, the controller which coordinates network activity, and finally the individual switches that implement actions specified by the controller.

2.2 SDN Components

SDN incorporates four components that facilitate development of a programmable network with physically-separate forwarding and control planes: the protocol, controller, network switch, and network application. There may be one or many switches communicating with one or many controllers, yet the protocol remains standardized and consistent throughout communication.

2.2.1 SDN Protocol.

A traditional switch's control plane communicates with the forwarding plane directly using proprietary communication as both planes are located on the same device. In order for different switch forwarding planes in a software-defined network to communicate with a controller, both a physical link to the controller, and a standard communication protocol are needed. The physical link may be a point-to-point connection, or an indirect connection through several hops. This communication exists within links labelled a, b, and c in Figure 2.1.

2.2.1.1 OpenFlow.

The de facto protocol in use today is OpenFlow [11]. OpenFlow is designed to decrease the barrier to entry in experimenting with network protocols, with the additional goal of preventing ossification of network infrastructure. OpenFlow allows protocol experimentation by allowing the network administrator to separate production traffic from experimental traffic. When a packet arrives at an OpenFlow compatible switch, the switch will either forward the packet based on current flow rules (established by a controller and programmed by an administrator), drop the packet, or will default to sending the packet to

9

the controller. When a controller receives a packet from an OpenFlow switch, the event is known as a "packet-in" event [9].

Table 2.1: OpenFlow v1.3.4 Match Fields

Field	Description
OXM_OF_ETH_DST	Ethernet destination address
OXM_OF_ETH_SRC	Ethernet source address
OXM_OF_ETH_TYPE	Ethernet type of the OpenFlow packet payload
OXM_OF_IP_PROTO	IPv4 or IPv6 protocol number
OXM_OF_IPV4_SRC	IPv4 source address
OXM_OF_IPV4_DST	IPv4 destination address
OXM_OF_IPV6_SRC	IPv6 source address
OXM_OF_IPV6_DST	IPv6 destination address
OXM_OF_TCP_SRC	TCP source port
OXM_OF_TCP_DST	TCP destination port
OXM_OF_UDP_SRC	UDP source port
OXM_OF_UDP_DST	UDP destination port

The application running on the controller dictates how to respond to a packet-in event. The controller application informs the switch using an event known as a "packet-out" event. Responses can include dropping the packet, forwarding the packet back to the switch with information about where the packet should go, or installing a rule on the originating switch. Each rule is known as a flow entry. A flow entry informs the switch about how to handle future packets that match a set of specified fields. Installing flow entries prevents the controller from intervening on every packet and thus prevents a bottleneck. Required available matching fields for a switch to be OpenFlow compliant are included in Table 2.1.

10

These matching fields are parsed by the Openflow switch. An OpenFlow switch may be capable of matching on additional fields other than the required list shown in Table 2.1.

Wildcards are available for certain matching fields. An example of using wildcard matching includes aggregate flows, which are flows installed in an Openflow switch that match on multiple IP addresses. Aggregate flows can dictate a switch to have a number of ingress ports forward to a specified egress destination.

Each flow entry can be created with an idle timeout period, where the flow entry is automatically removed if zero packets match the flow entry's headers for the duration of the timeout period. Flow entry statistics can also be gathered by the switch as another metric or input for programming. Available flow entry statistics depend on the network switch's capabilities and are included in Table 2.2. When a set of fields from an inbound packet matches a particular flow entry, a corresponding action is then applied to the packet. An action can be anything within the OpenFlow specifications and includes anything from packet-mangling to simply forwarding the packet to a specified port.

Table 2.2: OpenFlow v1.3.4 Available Flow Statistics

Statistic	Description
duration_sec	The time the flow has been alive in seconds
duration_nsec	The time the flow has been alive in nanoseconds beyond duration_sec
priority	The priority of the flow entry
idle_timeout	The number of seconds idle before expiration
hard_timeout	The number of seconds before expiration
packet_count	The number of packets that have traversed the flow
byte_count	The number of bytes that have traversed the flow

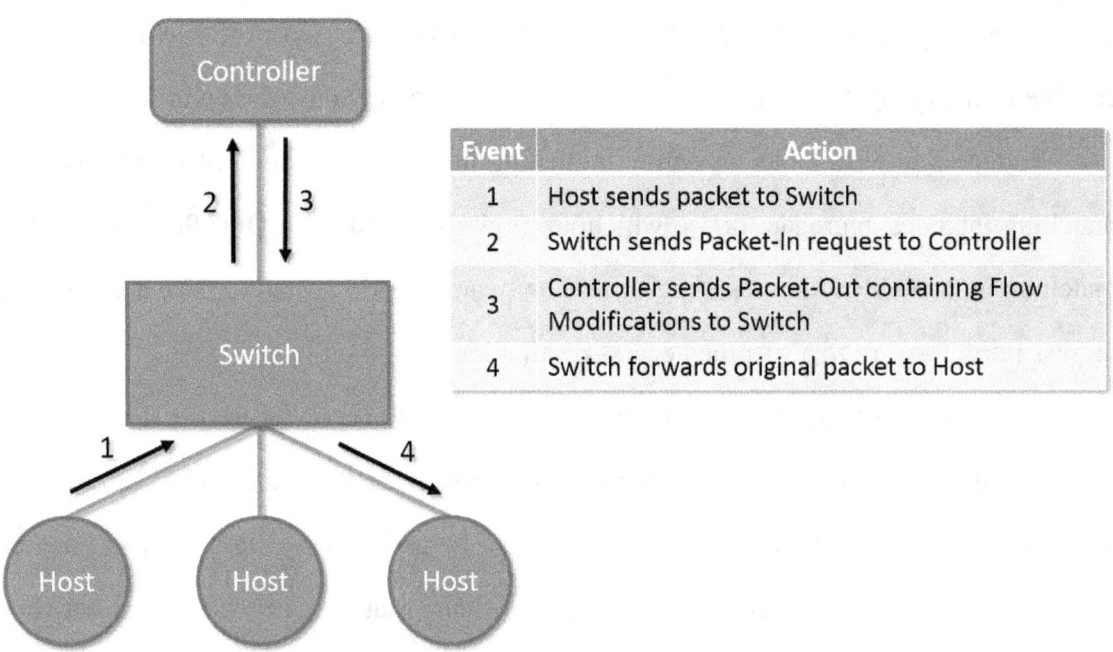

Event	Action
1	Host sends packet to Switch
2	Switch sends Packet-In request to Controller
3	Controller sends Packet-Out containing Flow Modifications to Switch
4	Switch forwards original packet to Host

Figure 2.2: SDN Packet-In and Packet-Out Events

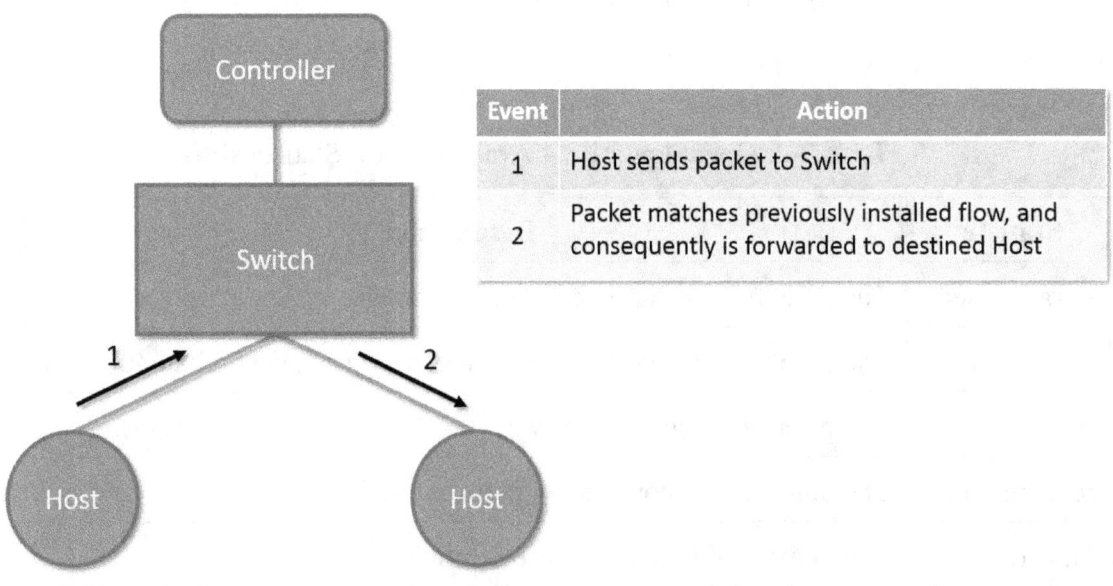

Event	Action
1	Host sends packet to Switch
2	Packet matches previously installed flow, and consequently is forwarded to destined Host

Figure 2.3: SDN Packet-In and Packet-Out Events

An example of the flow of communication is depicted in Figure 2.2. Within this figure, the left host creates a packet destined for the right host. The packet is sent to the switch. Without any flow rules installed, the switch does not know how to handle the packet because the packet matches no existing flow rule, and consequently the switch creates a packet-in event destined for the controller. The structure of the packet-in event is depicted in Figure 2.4 [12]. The packet-in event contains the entire Ethernet frame that was sent to the switch, should the controller need to parse it. The controller application then responds to the packet-in event and constructs a packet-out event destined for the switch. The structure of the packet-out event is depicted in Figure 2.5 [12]. The switch, responding to the packet-out event from the controller, forwards the original packet to the intended destination. If a flow is installed on the switch from the packet-out event, then future packets originating from the left host in Figure 2.2 that match the flow specification will forward to the right host without any communication with the controller (as depicted in Figure 2.3).

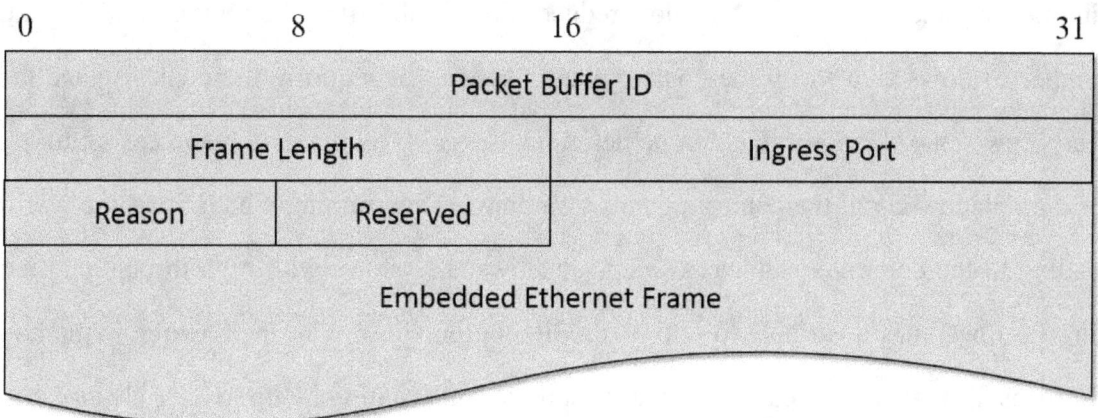

Figure 2.4: OpenFlow v1.0 Packet-In Structure

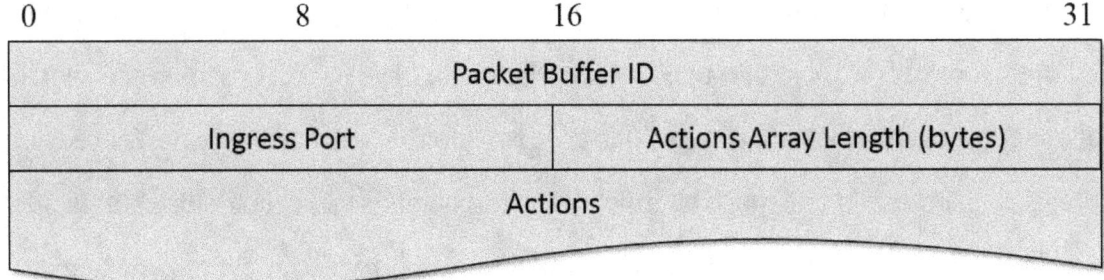

0	8	16	31

Figure 2.5: OpenFlow v1.0 Packet-Out Structure

2.2.1.2 DevoFlow.

Curtis et al. develop DevoFlow, another SDN protocol [13]. DevoFlow is designed in response to concerns about OpenFlow meeting the needs of high-performance networks. The granularity of OpenFlow control and visibility over flow entries requires a high rate of communication between the switch and the SDN controller, both for flow setups as well as for statistics-gathering. Such overhead limits the scalability of OpenFlow for high-performance networks. DevoFlow reduces the scalability concern by minimizing the number of flows created by the control plane. Instead of requring the control plane for every flow setup, all flows that can be determined locally by the data plane are set up by the data plane without the control plane's awareness. Determining which flows are worth sending to the controller, known as elephant flows (i.e., flows with high throughput and long life) becomes a variable to optimize. Simulation shows that the DevoFlow method improves throughput compared to equal-cost multi-path routing by up to 32% [13].

2.2.2 Controllers.

The network administrator programs the network control plane within the controller. The network controller is considered the network operating system; the controller provides an abstraction to the network topology much like an operating system abstracts the management of processes and memory. Through the controller, network applications provide the ability to both observe and manage network events. Several OpenFlow-compatible controller platforms are available.

Gude et al. created NOX, the first SDN controller [14]. NOX facilitates developing network applications, and thus satisfies the need of a network operating system. NOX runs on a sever, and contains a single network view that applications utilize to manage the state of the network. NOX works with the OpenFlow specification so that a standard communication protocol exists between the controller and network switches. Applications for NOX are written in Python or C++, with speed-critical core infrastructure written in C++.

Erickson created Beacon, the open-source OpenFlow controller written in Java [15]. Beacon makes OpenFlow applications both easy to develop and performance capable, as it bridges the gap between the ease of Python and the performance of C++. Beacon also allows OpenFlow applications to start and stop at runtime, giving the controller more of an operating system feel.

Zheng et al. propose an OpenFlow controller that overcomes the fact that NOX is a single-threaded system [16]. The controller, called Maestro, incorporates parallelism to address the concern that the controller could be a bottleneck. Maestro also takes advantage of batched socket write operations. Writing to the socket in batches requires less overhead than writing to the socket for each individual packet. Maestro also offers the ability to bind a worker thread to a core, which removes the overhead of moving code and data between cores, and results in increased throughput. Additionally, Maestro also allows binding an input flow request to a single working thread, keeping each request within one processor core, which reduces synchronization overhead.

A NOX controller with an input flow request rate of less than 20,000 requests per second (rps) results in an average delay of 2 ms. Increasing the request rate to the maximum throughput of NOX (i.e., 21,126 rps) increases the average delay to 17 ms. Increasing beyond this upper-bound results in packet buffering, causing TCP to slow down, resulting in an average delay of 4.11 seconds. This is due to the fact that the controller is only capable of utilizing one CPU core. When Maestro utilizes seven threads, an input rate of 630,000 rps results in an average delay of 76 ms. An input rate exceeding the maximum throughput of Maestro (i.e., 633,290 rps) causes an average delay of 163 ms. Under loads less than 16,000 rps the average delay is 2 ms. A NOX controller has a lower packet-handling delay for requests rates lower than 16,000 rps as compared to Maestro considering the overhead applied by Maestro to allow parallelism. Maestro's benefits are shown when the number of requests increases, allowing flexible scalability.

Tootoonchian et al. present cbench, a tool to quantify controller performance [17]. Cbench focuses on latency when a flow is first installed on a switch, as that is the likeliest source of a performance bottleneck. Cbench emulates OpenFlow switches which then send a custom number of OpenFlow packet-in messages. When a response arrives from the controller, cbench records the delay between request and response. In addition to cbench, Tootoonchian et al. creates a multi-threaded version of NOX known as NOX-MT to test NOX, Beacon, and Maestro using the cbench tool. Similar to Maestro, NOX-MT also performs batched socket write operations to minimize the overhead of writing to a socket. Unlike Maestro, which is implemented in Java, NOX-MT is a modification of NOX written in C++. NOX-MT also uses a modified malloc implementation that is designed specifically for multiprocessors. The use of a multiprocessor-aware malloc allows scalability within multi-core systems. When comparing maximum throughput, NOX-MT is capable of handling greater than one million rps when utilizing four CPU cores, while Maestro and Beacon both handle less than 500 thousand rps when using four CPU cores.

Floodlight is a self-proclaimed leader of the open source SDN controllers [18]. Written in Java, Floodlight introduces a modular loading system which simplifies updates and extensions. Floodlight is developed by an open community of developers including developers from Big Switch Networks [19]. Floodlight uses OpenFlowJ-Loxigen, a Java library generated by Loxigen [20]. Loxigen abstracts away low-level protocol details allowing an OpenFlow version to be easily defined. Any update to the OpenFlow version requires only a new Loxigen file defining the OpenFlow version through a common Application Programming Interface (API). The use of Loxigen by Floodlight prevents obsolescence due to OpenFlow version updates.

Lee et al. create Iris, a controller designed specifically to address the requirement of availability while allowing network scalability [21]. The architecture of IRIS allows multiple servers benefiting a controller cluster, using server resources on demand. Iris splits a network into independent "unit SDN networks" that are interoperable through a mechanism operated by a controller. Each unit SDN network has a scalable middleware the operates on top of the Openflow protocol, and can be controlled by a separate controller instance, providing horizontal scalability. Using cbench, Lee et al. model 100,000 hosts on an Intel server with a Xeon X5690 processor (having 3.47 GHz and 6 physical cores) and 64GB of RAM [17]. Analyzing flow processing performance demonstrates that the Iris controller outperforms Floodlight by a factor of 2.5 (i.e., the maximum number of flows maintainable is greater by a factor of 2.5). Floodlight is chosen for comparison as it is the most popular Java-based open-source Openflow controller [18].

While various research attempts to increase the efficiency of logically and physically centralized SDN controllers, other research has proposed physically distributing the controller while maintaining logical centralization. Tootoonchian et al. propose HyperFlow, an application designed for NOX controllers that creates a distributed event-based control plane for OpenFlow applications [22]. HyperFlow pushes state information to other controllers, allowing all controllers to have the same view of the network. Considering each controller has a uniform view of the network, each controller is locally able to make decisions for new flow requests. HyperFlow uses a publish/subscribe messaging paradigm to allow cross-controller communication. The publish/subscribe system is implemented using WheelFS, a distributed file system. Using WheelFS, channels between controllers are represented using directories, and messages are represented using files. HyperFlow polls directories to update changes made by other controllers.

Evaluation of HyperFlow performance is based on the window of inconsistency (i.e., the amount of time for controllers to agree on a view of the network) as load increases. The window of inconsistency is dependent on how fast HyperFlow and WheelFS are together capable of reading and writing files. HyperFlow is able to achieve a bounded window of inconsistency among controllers so long as fewer than 1000 link state changes occur per second (i.e., fewer than 1000 switches or hosts join or leave the network per second).

With the number of controller platforms increasing, it becomes necessary to compare various controllers. Shalimov et al. present an analysis of efficiencies of various open source SDN controller platforms using a new framework called hpcprobe [23]. The efficiency metric is determined by factors including performance (i.e., throughput and latency), scalability, reliability (measured through number of failures), and security (tested through the use of malformed Openflow packets). Hcprobe emulates an SDN environment, and allows custom tests through the use of its API written in Haskell. The SDN controllers tested include NOX, POX, Floodlight, Beacon, MuL, Maestro, and Ryu. In testing throughput, 32 switches are emulated with 10^5 hosts per switch. Beacon achieves the highest average number of flows per second as the number of threads increases. When testing reliability, MuL and Maestro drop controller-bound packet-in messages under the test load. When testing security through the use of a malformed Openflow header, Maestro and NOX crash upon receiving messages with incorrect length field values.

2.2.3 Switches.

Greenhalgh et al. describe two trends within network topologies: an increase in the proliferation of middle-boxes, and the commoditization of servers and switches [24].

Middle-boxes include any device sitting between switches, often providing transport-layer functions up to application-layer functions (e.g., web proxies, firewalls, intrusion detection/prevention, load balancers, etc.). The increase in switch capabilities combined with a reduction in cost leads to the consolidation of middle-box capabilities. This consolidation replaces middle-boxes with a single device: the programmable switch.

The benefits of consolidation include a reduction in equipment and maintenance costs, dynamic resource allocation, and an increased tolerance of failures. Rather than having separate hardware with dedicated (and thus rigid) functionality, the increased switch performance allows that functionality to be programmed dynamically into the switch at the network administrator's discretion, preventing the need for separate dedicated devices. Tolerance of failures comes from the fact that cheaper commodity hardware can be made redundant and easily hot-swapped. In order for switches to contribute to a software-defined network, they must be compliant with the protocol utilized by the controllers. The requirement for SDN-capable switches limits the availability of switches to those compliant with the OpenFlow specification.

2.2.4 Applications.

An SDN application runs locally on the controller and dictates either proactively or reactively how forwarding tables are populated. An emulated environment for SDN applications allows a network administrator to test whether a network application fulfills specific requirements. Additionally, an emulated environment allows a network administrator to test the SDN application before installing it onto production controllers. Lantz et al. developed Mininet, a system that generates an emulated environment which allows rapid prototyping for large networks [25]. Mininet addresses the problem of a high barrier to entry for designing a new network architecture. Mininet is designed to be flexible, deployable, interactive, scalable, realistic (i.e., the behavior observed in the emulated

19

environment should represent what would be expected in a production environment), and shareable (i.e., easily encapsulated and installed).

Mininet offers a Command-Line Interface (CLI) that allows the user to easily create an arbitrary network with virtual hosts attached to any number of OpenFlow switches and a controller of any type. Mininet also contains a Python API which allows custom scripting for experimentation. One notable limitation of Mininet is the fact that software forwarding speeds cannot match the $O(1)$ forwarding time of hardware lookup tables, which may skew packet forwarding-rate statistics.

Gupta et al. developed a simulation tool known as fs-sdn, which is designed for greater scalability than Mininet [26]. Rather than basing the simulated network abstraction around each packet, fs-sdn's unit of measure is a flowlet. A flowlet represents the number of flows emitted over a period of time (e.g., within a period of 150 milliseconds, 1 or more packets traversed the emulated device). The higher level of abstraction allows greater speed and accuracy than current packet-level simulators (e.g., ns-2 and ns-3) [27][28]. Results show that fs-sdn remains consistent across various topologies and rate configurations, while Mininet loses accuracy as both load increases from 10 Mb/s to 100 Mb/s, and with an increase in the number of nodes within the emulated network. Accuracy is determined by comparing byte, packet, and flow counts recorded by each platform.

2.3 SDN Growth

Increasing interest in SDN capabilities among SDN researchers as well as growing applications of SDN environments emphasizes the importance of security as its widespread adoption presents a more lucrative target for network attackers. Google is one example of a major corporation taking advantage of SDN capabilities. Jain et al. present B4, a software-defined network connecting Google's data centers together across continents [29][30]. Specific needs inclusive to high bandwidth, dynamic traffic demands, and full control over edge servers, led to the choice in adopting Openflow with an SDN environment

20

to accomplish the needs of B4. In addition to Google, the SDN market is expected to grow beyond $35 billion by April 2018 [31]. The rate of growth is shown by the fact that SDN technology sales have increased from $10 million in 2007 to $252 million in 2012 [31].

2.4 SDN Security Concerns

Considering the age of SDN, concerns exist regarding the security implications imposed by the transformation of an entire networking paradigm. Klöti et al. use STRIDE and attack tree modeling methods to evaluate the security of an emulated SDN network [32][33][34]. When testing the potential for denial of service, the POX-controlled MiniNet simulated SDN environment is utilized with an increased timeout value to allow greater flow entries. The number of lost packets increases consistently with an increasing flow timeout value (and consequently an increase in table overflows). When testing for information disclosure, Klöti et al. compare the time it takes for an initial connection of a server to the time it takes for a subsequent connection with the same network server. A drastic difference in delays denotes a new flow entry, which suggests no other client already has an established session with the service of interest. This type of information disclosure only exists when flow aggregation is utilized. Klöti et al. suggest rate limiting, flow aggregation, attack detection, and access control mechanisms to mitigate any Denial of Service (DoS) vulnerabilities. Klöti et al. also suggest proactive strategies inclusive to randomizing response times as well as attack detection methods in order to mitigate potential information disclosure vulnerabilities.

Scott-Hayward et al. present an in-depth survey on research related to security implications of SDN [35]. Within the survey, two schools of thought regarding security in SDN environments exist. The first states that the programmability of the new technology allows a more secure environment as well as a centralized view of the network security posture. The second states that the centrality and other new features of SDN expose the network to a new vector of attacks.

Shin and Gu propose a feasibility study in attacking SDN environments [36]. The attacks include fingerprinting an SDN environment as well as using a DoS attack against an SDN environment. Fingerprinting involves the use of SDN Scanner, which changes the header field in two packets, and collects the RTT values for these packets. From the RTT values, Shin and Gu can determine whether an externally-located network is an SDN environment with a success rate of 85.7%. The DoS analysis tests against Open VSwitch installed on a separate Linux host, with a switch flow table capacity of 1500 flows entries. Shin and Gu show with the emulated SDN environment that it takes ten seconds with 200 packets per second at a bandwidth of 0.75 Mbps to effectively consume the resources of the control plane and the data plane.

Documented vulnerabilities already exist with current SDN controllers. Dover presents two, both targetting the Floodlight SDN controller [37][38]. The first vulnerability is a switch table vulnerability, where malformed Openflow messages allow an attacker to cause full CPU utilization on the controller, ultimately denying controller functionality. The second vulnerability is another DoS vulnerability that involves tearing down the connection between the targeted switch and controller. Without communication between the controller and the switch, the implementation of the switch dictates whether the network switch falls back to traditional forwarding techniques or follows another response to the DoS attack.

Diego et al. present seven threat vectors regarding the introduction of the SDN environment [39]. The first threat vector includes forged or faked traffic flows, which target switches and controllers in an SDN environment. The second threat vector includes attacks on vulnerabilities specific to Openflow network switches. The third threat vector is an attack on the control plane communication (i.e., the link between the network switch and the SDN controller). The fourth threat vector includes vulnerabilities existing in the SDN controller, given the controller is software potentially consisting of numerous faults. The fifth threat vector is the lack of verification between the controller and the management applications written on them. The sixth threat vector includes vulnerabilities within administrative stations that exist within an exploitable network. The final threat vector is the lack of trusted resources for forensics and remediation for when an attack is detected.

2.5 SDN Security Applications

Numerous applications have been created that apply the SDN architecture towards providing security to an existing network. The logically centralized nature of the control plane allows an additional component that is taken advantage of through security applications: a view of the entire network. Not only does SDN allow a singular view of the entire network, but SDN also allows a controller to change that view, lowering the barrier to creativity in network programming. Security applications include anomaly detection, Distributed Denial of Service (DDoS) prevention, Intrusion Prevention System (IPS), modular security, and a moving target defense.

2.5.1 Traffic Anomaly Detection using SDN.

Mehdi et al. utilize SDN to revisit Anomaly Detection Systems (ADS), a security application that was explored "rather unsuccessfully" in the past [40]. Mehdi et al. differentiate between the network core and the network edge. The network core consists of hardware that makes up the backbone of the Internet. The core does not directly touch the end devices, but rather facilitates the end device communication by connecting sub-networks to each other. The network edge includes switching hardware that directly connects to an endpoint device or endpoint network. The method explained in [40] involves moving four different ADS algorithms from the network core to the network edge. Before SDN, this was difficult as switching hardware ran proprietary software that was not modifiable. By replacing proprietary switches at the network edge with openly-programmable switches adherent to the OpenFlow specification, ADS can be moved closer to the anomalous sources (i.e., the network edge).

23

Moving ADS closer to the anomalous sources offers different magnitudes of benefits depending on the ADS algorithm implemented. One ADS algorithm implemented was the Threshold Random Walk with Credit Based rate limiting (TRW-CB) algorithm. TRW-CB is based on the notion that connection attempts are more likely to succeed when originating from benign hosts rather than from malicious hosts. Implementing the TRW-CB algorithm at the network edge (using a home network dataset) with SDN was shown to achieve a 90% accuracy of detection, with a false positive rate of 0% to 4%. Moving this algorithm to the network core (using an Internet Service Provider (ISP) dataset) resulted in an accuracy of 85% detection rate with a false positive rate of 11%.

2.5.2 DDoS Flooding Attack Detection.

Braga et al. use the per-flow statistics already stored by OpenFlow compliant switches to detect a DDoS [41]. The detection process involves three modules: the flow collector module, the feature extractor module, and the classifier module. The flow collector module requests flow statistics from OpenFlow switches at an interval specified by the controller application. Using flow statistics rather than packet header information decreases overhead. The feature extractor module grabs data relevant in detecting a DDoS flooding attack. The classifier module uses a Self-Organizing Map (SOM), a type of neural network, to classify whether the polled data consists of a DDoS flooding attack.

The features worth extracting from flow statistics include the Average Packets per flow (APf), the Average Bytes per flow (ABf), the Average Duration of the flow (ADf) (i.e., the flow's lifetime), the Percentage of Pair-flows (PPf), the Growth rate of Single-flows (GSf), and the Growth rate of Different Ports (GDP). All flow statistics are natively logged by OpenFlow switches, preventing the need to install custom software for traffic statistic logging. Using flow statistics resulted in a detection rate greater than 98%, while taking a CPU time of 154 seconds to extract 6 flow features on a 1.8 GHz, dual core CPU with 2 GB of RAM. Feature extraction methods using packet header information required

9 features (taking a CPU time of 237 seconds to extract on a 2.66 GHz, dual core CPU with 3.5 GB of RAM).

2.5.3 *Intrusion Prevention System with SDN.*

Xing et al. propose an IPS that uses OpenFlow to perform countermeasures on flows identified by Snort's Intrusion Detection System (IDS) engine [42]. Working in tandem with an iptables firewall, an IDS effectively becomes an IPS. Such an IPS system suffers from latency issues, accuracy issues (in terms of the number of false positives), and flexibility. While [42] claims to compare SnortFlow with an iptables-based IPS, no data is shown directly comparing these two IPS systems.

2.5.4 *Modular Security Services.*

Shin et al. present FRESCO, an OpenFlow application development framework that utilizes standardized modules to facilitate security detection and mitigation development [43]. FRESCO overcomes an information deficiency challenge through the use of a database module (FRESCO-DB) that simplifies storage of key state tracking information. Such state tracking information is normally not collected by OpenFlow controllers, yet is often required to develop security applications. Example data that exists in the FRESCO-DB module includes a TCP connection status as well as IP reputation. Replicating open-source network security applications using FRESCO results in an order of magnitude fewer lines of code. FRESCO also deploys garbage collection, which checks if an OpenFlow switch is nearing the capacity of flow entries for the switch, and evicts the least active flow. Garbage collection in FRESCO minimizes resource utilization by reducing the number of flow entries within each OpenFlow switch. One drawback to FRESCO exists in the setup time required to install flows. FRESCO applications require additional setup time to install a flow entry ranging from 0.5 milliseconds to 10.9 milliseconds more than the time it takes for the default NOX controller to install a flow entry. The additional setup time was credited to using an emulated environment running on a virtual machine, but this theory was not tested.

2.5.5 *Moving Target Defense using SDN.*

Jafarian et al. take advantage of the fact that the controller monitors and manipulates the entire network from a central vantage point, and use the centralized controller to provide a moving target defense against network attacks [44]. The technique, called OpenFlow Random Host Mutation (OF-RHM), assigns a random virtual Internet Protocol address (vIP) for each end hosts within a network, while maintaining a map of vIP to a real Internet Protocol addresses (rIP) handled by the controller. The controller then installs flows necessary for OpenFlow switches to forward a given packet to the correct destination. Each connection is associated with an installed flow, allowing a new vIP to be assigned to a host in the middle of an established session between two hosts. Performing an Nmap scan targeting a class B network with 2^{16} hosts results in a list of hosts believed to be active. Performing multiple Nmap scans reveals that not more than 1% of the vIP addresses obtained from the initial scan remain consistent. Analysis on worm propagation effectiveness reveals that OF-RHM saves up to 90% of hosts against being caught through known worm propagation techniques (e.g., divide-and-conquer worms that cooperate with each other to ensure no host is scanned twice).

2.6 SDN and Latency Analysis

Within the Openflow protocol, two fields exist dictating the length of time for a switch's flow entry to remain active. These fields include the flow inactivity timeout, and the flow hard timeout values. When a controller sends an Openflow message to a switch, informing the switch of a flow action, this Openflow message also contains both the inactivity and hard timeout values. An inactivity timeout value directs a flow to expire when an amount of time passes without any packet matching the flow. A hard timeout value directs a flow to expire after an amount of time passes since the flow's creation, regardless of matching packet traffic. Both the inactivity and hard timeout values can be set to zero, indicating an infinite value.

Zarek et al. explore the effect these timeout values have on performance measured through the flow miss rate, and table occupancy [45]. Zarek et al. find that an increase in the flow timeout value allows the flow to reside within the switch longer, decreasing the miss rate exponentially, while also growing the table size linearly. Zarek et al. also observe an optimal point at which any increase in the flow timeout value results in negligible miss rate benefit, while also unnecessarily adding to the table occupancy, leading to a potential upper-limit in the number of flows. Zarek et al. propose two flow table management methods that combine optimal flow timeout values with explicit messages from the controller. These explicity messages request the switch to evict specified flow table entries. Applying these two flow table management methods to two different data sets, Zarek et al. discover that the use of the explicit eviction method and optimal flow timeout values result in large savings in flow table size as the number of table entries increases into the tens of thousands.

Vishnoi et al. propose SmartTime, an Openflow controller that also uses varying timeout values combined with explicit flow eviction messages to optimize controller load as well as internal switch memory [46]. SmartTime's adaptive idle timeout strategy is applied to determine the effect on both flow table misses and flow entry drops. The adaptive strategy improved the current static inactivity timeout policy's performance in terms of both flow table misses and dropped flow entries.

Kim et al. attempt to help controller scalability by recognizing that reducing the number of flow setup requests reduces the load placed on controllers [47]. The method in which Kim et al. reduce the number of flow setup requests includes minimizing the number of flow entry evictions. Dynamically setting flow timeout values allows common flows to have a longer timeout value, reducing the flow's chance of eviction. With fewer evictions, the Openflow switch makes fewer flow requests, reducing the load on the controller. The dynamic flow timeout algorithm reduces the number of packets sent to the controller by 9.9%.

2.7   SDN Applied to Network Security

The research presented in this thesis involves gathering information about an unknown network, and then from that reconnaissance, if the network is identified as an SDN environment, determining the controller supporting the SDN environment. The information gathered includes features of each controller inclusive to both inactivity and hard flow timeout values. Populating a table, an increasing number of controller features allows unique identification of both the controller instance as well as potentially the version of the controller software. From discovering the type of controller active within the SDN environment, documented vulnerabilities pertaining to the controller may exist that allow an attacker unmitigated access to a network.

III. Methodology

This chapter presents the methodology for evaluating the process of fingerprinting the SDN controller. Section 3.1 lists the specific questions that are addressed by each experiment. Section 3.2 describes the objective and hypothesis of this thesis, while Section 3.3 explains the approach to accomplish the stated objective. Finally, Section 3.4 describes each experiment in detail, including the assumptions, parameters, hypothesis, and design for each experiment.

3.1 Problem Definition

Traditional networks consist of switches and routers that contain static protocols, removing any ability to create custom network protocols. In order for a network administrator to define a custom routing protocol, they are not able to considering the device's firmware is static. The network administrator is then left to requesting the device vendor for a specific feature [9]. SDN moves the decision-making logic within a network from a distributed set of switches and routers, to a logically-centralized controller. The SDN controller communicates to each network switch through the OpenFlow protocol.

Shifting from a traditional network to an SDN environment brings changes observable by a client in the SDN environment. While the end result allows a packet to reach an intended destination, the following process is designed to determine the underlying controller that is controlling the SDN environment. Identifying the controller (a process referred to as fingerprinting) allows a malicious client to tailor a specific attack and can lead to a discovery of vulnerabilities to exploit. Additionally, experiments show that a client can obtain information programmed into the controller. Determining what information an attacker can glean contributes to a broader understanding of attacker capabilities available in an SDN environment. Further, any security postures that depend on the assumption of an attacker's lack of information are shown invalid.

3.2 Goals and Hypothesis

The objective of this thesis is to develop and evaluate a process for uniquely identifying the controller supporting an SDN environment. The proposed process includes extracting features from a new SDN controller and adding the new feature set to a table of features. A network client can then retrieve features of an unknown SDN controller and query the retrieved features against the table of known controllers. The first goal of this research is to construct a set of features extensive enough to uniquely identify each known SDN controller, and demonstrate that the table of features reliably identifies each SDN controller. The next goal includes ensuring that each feature is obtainable by a client connected to the SDN environment. With both of these goals achieved, a client will successfully be able to add a new SDN controller to a feature table, and subsequently identify that controller in an SDN environment.

It is hypothesized that a process can be created that adds each new SDN controller into a table of SDN controller features, and that this table can be used to identify an unknown SDN controller discovered in an SDN environment by a client connected to that environment.

3.3 Approach

This section provides a high-level view of the process for extracting and using SDN features to fingerprint an unknown SDN controller, and describes how this approach fulfills the goals mentioned in Section 3.2.

3.3.1 Overview.

The abstract design of the fingerprinting process is shown in Figure 3.1. The right box in Figure 3.1 shows an unknown controller in an SDN environment. The controller represents a company or entity using a controller that adheres to the OpenFlow specifications and is not in any table of features. As a client seeking to fingerprint the known controller of an SDN environment (shown as a red circle), the client downloads a copy of the publicly available SDN controller and adds the SDN controller's features to the table of features through feature extraction. Next, the client observes the features of the unknown SDN controller and determines whether they match a set of features in the table.

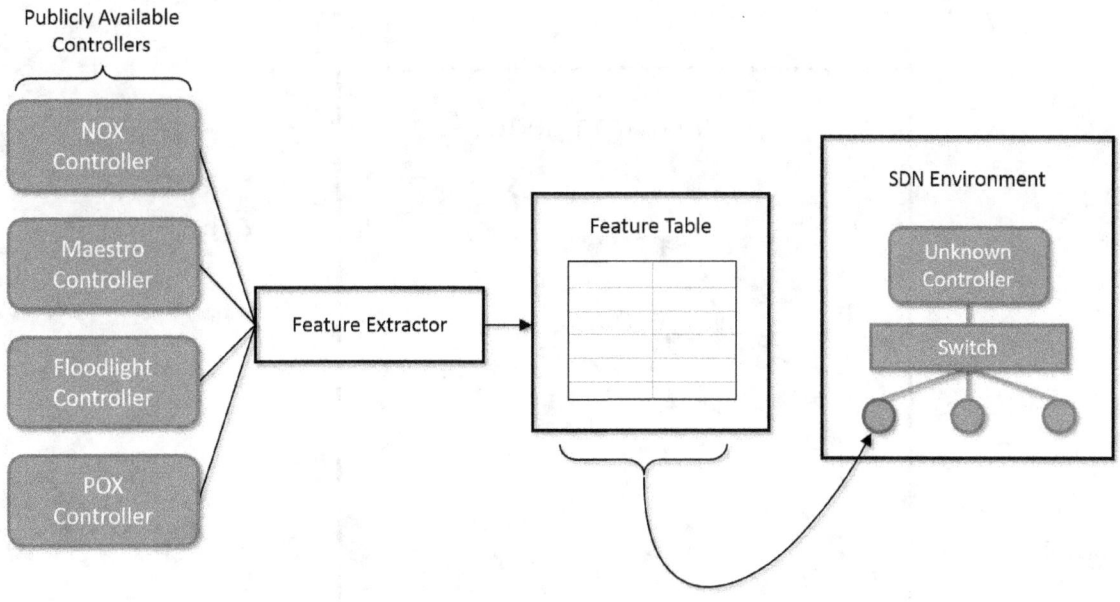

Figure 3.1: SDN Controller Fingerprinting Process

With access to a new SDN controller, an administrator can extract any and all features by examining the source code (provided the SDN controller's source code is available), reverse engineering the SDN controller, or using the SDN controller in a network environment and discovering features. It is important to note that the set of features observable by the client is limited compared to the set of features extractable by an administrator of that SDN controller. Emphasis is placed on the fact that administrators can extract features, while clients can observe features. A network administrator, as depicted in Figure 3.2, has complete control over the SDN controller, while the client can only communicate with the SDN controller through the provided network interface. The client's limited visibility of the SDN controller provides a requirement that must be fulfilled by each feature extracted into the table: each feature must be observable by a client connected to the network. If a feature inserted into the table is not observable by a client, then that feature cannot be used by the client to uniquely identify the SDN controller.

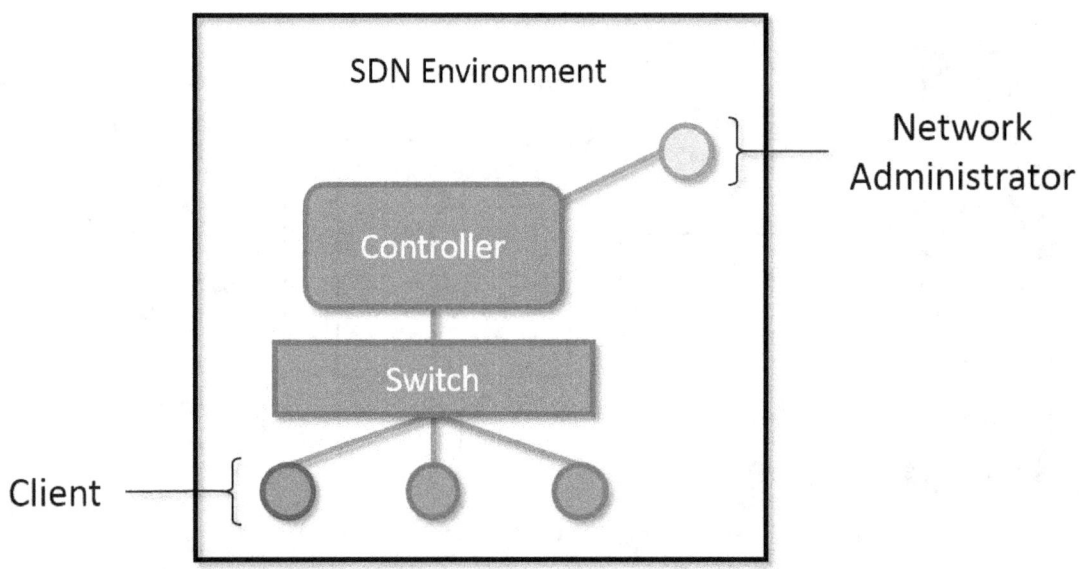

Figure 3.2: Client versus Administrator SDN Controller Visibility

During the feature extraction process, a client assumes the role of an administrator by downloading a copy of publicly available SDN controllers. By assuming the role of an administrator, the client is able to extract any feature and populate the feature table. After extracting the features of a new SDN controller, the client can then use the feature table while connected to an SDN environment and test whether the observed features of the environment's controller match the previously extracted features.

3.3.2 Features.

The choice of features is based on the principle of observing the response after any provided input. Figure 3.3 shows the OpenFlow protocol communication, and provides locations of where features can be extracted within the protocol. The left device in Figure 3.3 represents an OpenFlow switch, however the SDN controller does not distinguish between host and switch, so long as the device is communicating using the OpenFlow protocol. Point A represents the time just after the TCP 3-way handshake has concluded. Observing what the controller communicates immediately after point A may reveal unique messages that constitute a unique feature of the controller. An example feature at Point A is whether the controller begins sending the symmetric OpenFlow Hello packet, or if the SDN controller waits for a received OpenFlow Hello packet first. Point B represents the time just after the client sends the OpenFlow Hello packet. An example of an observable feature at point B includes the very next packet received from the SDN controller, which may be an OpenFlow Feature Request packet, adhering to the OpenFlow specification, or some other packet depending on the SDN controller implementation. The very next input provided by the client is a response to the OpenFlow Feature Request packet. Point C represents the time just after the OpenFlow Feature response is transmitted. Any future input by the client creates a new point in Figure 3.3, and thus adds a new feature to the feature table.

The list of features used for the current SDN controller fingerprint process is shown in Table 3.1. The time column represents the most recent action completed by the client, while the feature column shows the response observable from the SDN controller. A time of "any" signifies that the feature can be collected at any time during the connection between the client and controller, and does not follow any event listed in Figure 3.3. Adding a new SDN controller to the feature table may require finding new features that distinguish the new SDN controller from the SDN controllers already within the table. Finding a new

33

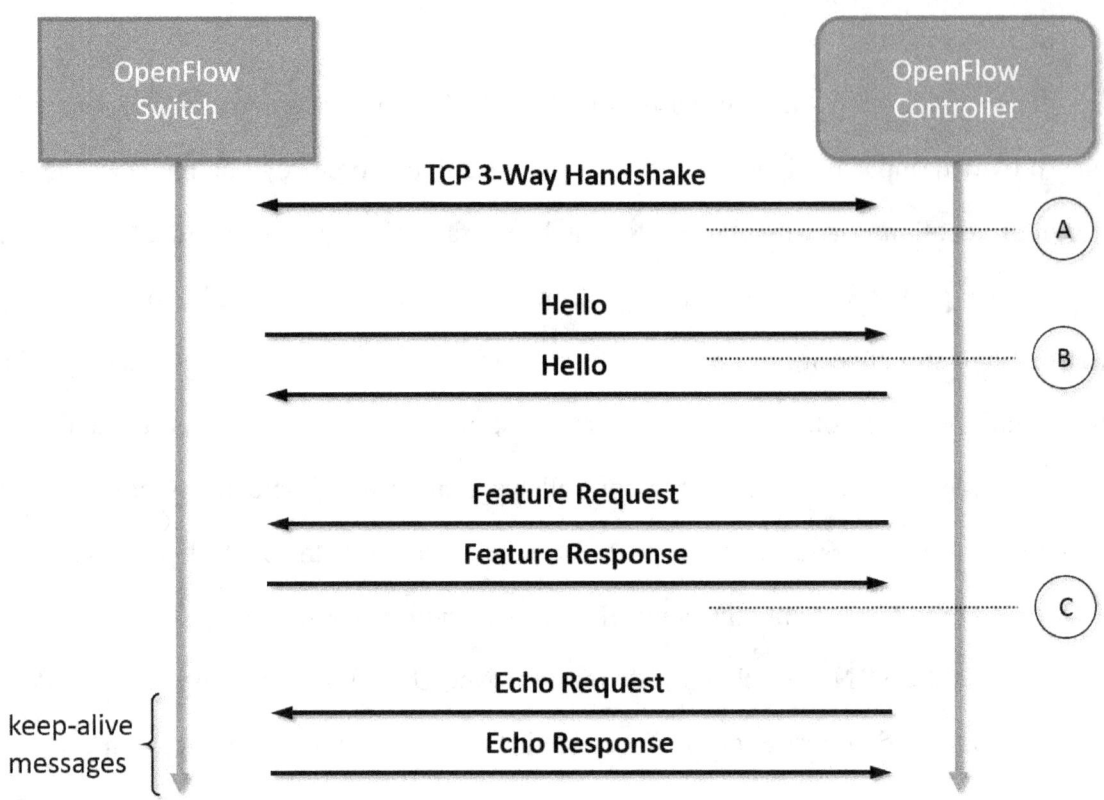

Figure 3.3: OpenFlow Protocol Feature Extraction Points

feature requires that the new feature's value be determined for each SDN controller already in the table. If a feature is obtainable by one SDN controller, but not another, then that lack of availability of a feature serves as identifiable information that is still populated within the feature table. Section 3.4 contains experiments verifying each feature as observable from the perspective of a client connected to the SDN environment.

Table 3.1: List of Features and Collection Times

#	Time	Feature
1	Any	Default Flow Inactivity-Timeout Period
2	Any	Default Flow Hard-Timeout Period
3	After TCP Handshake	Does the SDN controller transmit an OF Hello packet?
4	After TCP Handshake	Does the SDN controller transmit an Echo Request?
5	After TCP Handshake	Does the SDN controller transmit a Feature Request?
6	After Client Sends Hello	The set of packets sent by the SDN controller
7	After Client Sends Feature Response	The set of packets sent by the SDN controller

3.4 Experiments

This section provides specific details regarding each experiment conducted throughout this thesis. Each experiment begins with the reasoning for the experiment, a list of assumptions pertaining to the experiment, the parameters, the hypothesis, and finally the experimental design. Experiment 1 demonstrates that a client can accurately determine whether the connected network environment is a traditional environment or an SDN environment. Experiment 2 verifies that the first feature, the SDN controller's flow inactivity timeout period, can be accurately determined by a client connected the SDN environment. Experiment 3 verifies that the second feature, the SDN controller's hard timeout period, can be accurately determined as a client connected to the SDN environment. Experiment 4 shows that these features can combine to form a table of SDN controller features able to fingerprint the controller software.

3.4.1 Experiment 1: Verify SDN Environment.

The goal of Experiment 1 is to determine whether a client within a network environment can accurately determine if that network environment is an SDN environment or a traditional environment. Experiment 1 is required for fingerprinting the SDN controller

35

as it is the first step of discovering whether the client is in an SDN environment and thus whether an SDN controller exists. Determining the network environment, as well as determining the specific controller supporting the environment (if that environment is determined to be an SDN environment) allows the client to glean information that increases the viability of a network attack.

3.4.1.1   Experiment 1 Background.

As a client connected to a computer network with an unknown architecture (i.e., the client does not know whether the network is a traditional or an SDN environment) it may be possible to distinguish the network as an SDN environment. Considering each packet incurs a packet-in event (see Section 2.2.1 for information about packet-in events) from the switch to the controller before a flow rule becomes installed, latency for the first packet exists that is not observed by subsequent packets of similar type (i.e., subsequent packets that match the same flow rule). This latency is observed by comparing the path of a ping (i.e., Internet Control Message Protocol (ICMP) echo) packet in Figure 3.4 with the path of a subsequent ping packet in Figure 3.5.

Figure 3.4 shows a ping request packet created by Host A, destined for Host B, sent to the network switch. The switch, having never installed a flow for the ICMP packet, sends a packet-in request to the controller, which responds with a packet-out event sent to the switch. The packet-out event contains a flow modification request from the controller telling the switch to install a flow based on the initial ping request created by Host A. The packet-out event also tells the switch to forward the ping packet originating from Host A to the port occupied by Host B. At this point, all ping request packets matching the initially transmitted ping request packet (i.e., originating from Host A, at Host A's current port on the switch, destined to Host B), will forward to Host B's port on the switch without any negotiation with the controller. Host B responds with a ping response packet destined for

Host A, which repeats the process of the network switch negotiating with the controller and installing another flow.

Figure 3.5 shows a second ping request created by Host A, destined for Host B. This process is different than the process shown in Figure 3.4 because flow entries in the switch are already installed, so no packet-in events are required for the switch to correctly forward the ping packets. The difference in latencies between the processes shown in Figures 3.4 and 3.5 can be calculated from a client by recording a ping packet's RTT, and is consequently observable to a client connected to the SDN environment.

The difference in latency between the first packet and subsequent packets is not observed for traditional networks as there is no flow-installation process. Without a flow-installation process, both initial and subsequent ping requests and responses are treated as shown in Figure 3.5. Because of the latency difference between SDN and traditional networks, a distinction can be gleaned from analyzing packet latencies. This distinction can also be used to determine whether a flow is still active within the switch, as added latency is a result of a non-existent flow. Experiments 2 and 3 rely on this process of determining whether a flow is still active within the switch.

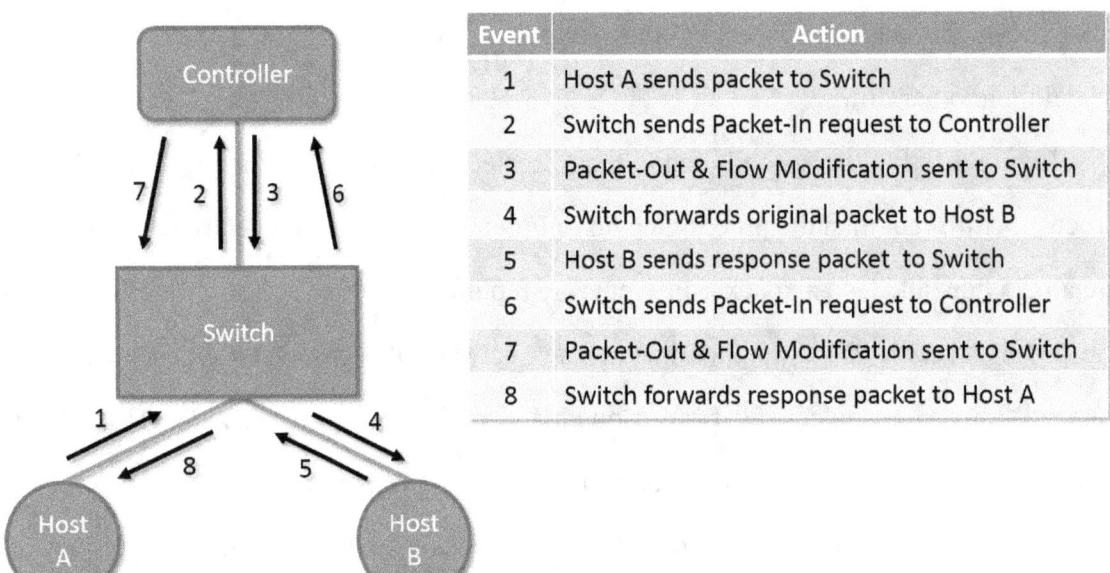

Event	Action
1	Host A sends packet to Switch
2	Switch sends Packet-In request to Controller
3	Packet-Out & Flow Modification sent to Switch
4	Switch forwards original packet to Host B
5	Host B sends response packet to Switch
6	Switch sends Packet-In request to Controller
7	Packet-Out & Flow Modification sent to Switch
8	Switch forwards response packet to Host A

Figure 3.4: Path of an Initial ICMP Echo Request and Response

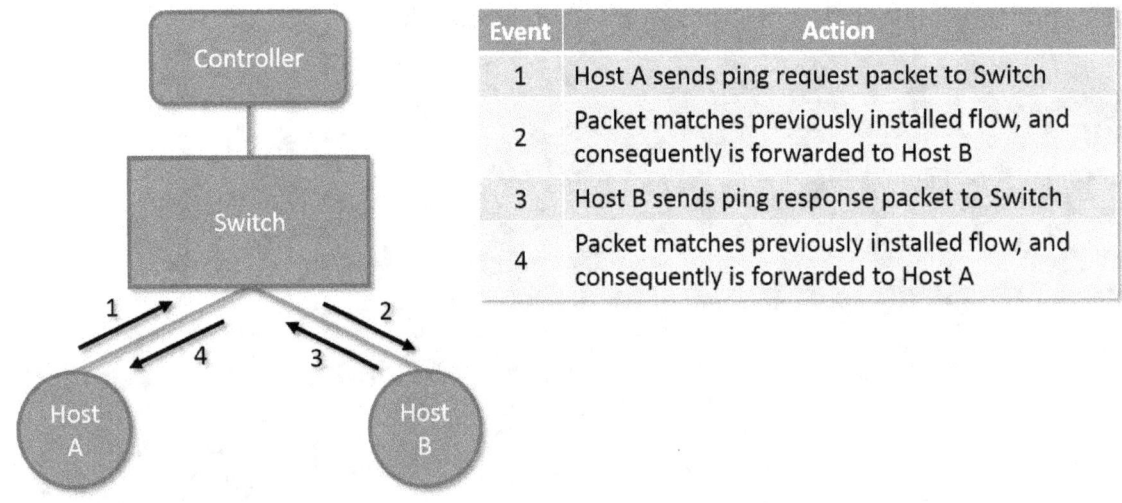

Event	Action
1	Host A sends ping request packet to Switch
2	Packet matches previously installed flow, and consequently is forwarded to Host B
3	Host B sends ping response packet to Switch
4	Packet matches previously installed flow, and consequently is forwarded to Host A

Figure 3.5: Path of a Subsequent ICMP Echo Request and Response

3.4.1.2 Assumptions.

The assumptions for Experiment 1 are as follows:

1. The client has access to a port on a switch connected to the SDN environment.

2. ICMP echo packets are not blocked by the sending, receiving, or intermediary node.

3. Aggregate flows are not installed on the switch.

4. Aggregate flows are not used by the SDN controller for communication between Host A and Host B.

5. The client knows the IP address of a live host that responds to ICMP echo requests.

3.4.1.3 Parameters.

Table 3.2 shows the parameters which define the treatment levels for both the simulated MiniNet test environment and the emulated test environment. The SDN controller parameter is none when the environment is a traditional environment because no controller is present. The emulated test environment has fewer SDN controllers because the hardware switch is compatible only with OpenFlow version 1.3, and fewer open-source SDN controllers exist that support this version. Table 3.3 shows the various parameters held constant to reduce the number of covariates. The response metrics include the ping RTT and the difference between the first ping RTT and the second ping RTT. The elapsed runtime for determining the predicted inactivity timeout value is also recorded. Recording the runtime allows future research to compare the speed in determining the inactivity timeout value, if a supposed faster process needs comparing.

3.4.1.4 Hypothesis.

The expected output for Experiment 1 will show two things: first, that within an SDN environment, the first ping latency is significantly greater than the second ping latency, and

Table 3.2: Experiment 1 Parameters

Test Environment	Parameter	Setting
Simulated (MiniNet)	Environment	SDN
		Traditional
	SDN Controller	NOX
		POX
		Beacon
		Maestro
		Floodlight
		NodeFlow
		OpenDaylight
		None
Emulated (Hardware)	Environment	SDN
		Traditional
	SDN Controller	Ryu
		Iris
		None

second, that this difference in latencies is significantly greater in SDN environments than in traditional environments.

3.4.1.5 *Experiment Design.*

Experiment 1 uses Mininet to simulate the SDN environment [25]. Within Mininet, a simple network consisting of two hosts connected to one switch with one controller (the control parameter) is used [14]. The simple network is depicted in Figure 3.6. Within Figure 3.6, the network switch, Hosts A and B are all Linux kernel version 3.8.0 emulated in a MiniNet virtual machine. The network switch is compatible with the OpenFlow

Table 3.3: Experiment 1 Parameters Held Constant

Parameter	Value
Number of Clients	2
Number of Switches	1
Number of Controllers	1
Number of Hops between Host A and Host B	2

standards [11]. The controller is hosted on a separate virtual machine running Ubuntu version 12.04 with Linux kernel version 3.8.0. Considering Mininet requires the use of an SDN environment, the traditional environment was constructed using two Ubuntu 12.04 virtual machines connected by VMWare's virtual network interfaces.

A bash script runs on Host A, pinging Host B three times with a one second delay between each ping. After three pings, the script waits thirteen seconds, and repeats 100 times. A pause of thirteen seconds is chosen to allow sufficient time for flow entries within the switch to expire due to inactivity. The default inactivity timeout for flow entries is dependent on the implementation of the controller, and thirteen seconds is greater than the default inactivity timeout available for every controller tested. While the bash script executes, tcpdump is running in the background on Host A, recording timestamps for each ping transmission and response. The first ping is sent to record a flow-setup time, as this ping will incur a packet-in and packet-out delay between the switch and controller. The second ping is sent to record a ping traversal time after a flow setup occurs. The third ping is used for a cursory comparison with the second ping and starts the thirteen-second wait period. When the flow entry expires due to inactivity, the very next ping requires a new flow installation for the switch. Repeating the 3-ping process 100 times is done to increase the sample size.

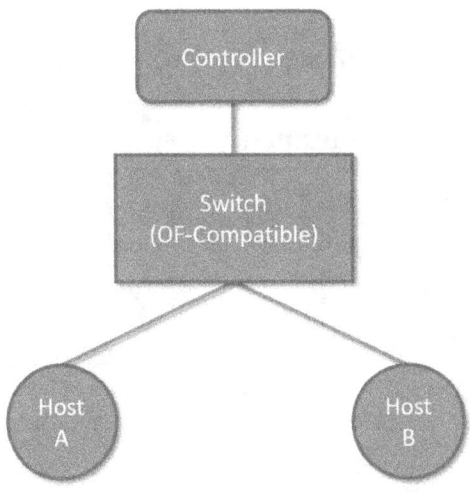

Figure 3.6: Mininet Simple Network

It is important to note that tcpdump is required for an accurate reading of timestamps. The ping application informs the user of the RTT for each ping request, however this RTT does not accurately reflect the moment an ICMP echo request packet was placed on the wire to the moment an ICMP echo reply packet was received on the wire. The ping application sets a timestamp as soon as a ping request is constructed by the application, but after this timestamp is recorded, Address Resolution Protocol (ARP) packets are sent and received to determine the next-hop path for the ping packet. These ARP packets incur delays that are included in the ping application's calculation of RTT, yet are of no interest for Experiment 1. Using tcpdump effectively allows the experimenter to record more precisely when the ping request is placed on the wire by observing when an ICMP echo packet is transmitted.

The tcpdump output is parsed using a python script that takes the difference between each ICMP echo request timestamp and the corresponding ICMP echo response timestamp. The python script outputs this difference, along with the corresponding ping number (first, second, or third) in a comma-separated values (csv) file that can be analyzed using R statistical programming. Multiple csv files are generated (one for each controller used, as well as one for a traditional environment not containing a controller) and are compared using the R programming language.

Experiment 1 is also conducted on separate hardware to compare real-world results against the simulation in MiniNet. The structure of the new environment is similar to the MiniNet structure, in that there are 2 hosts, 1 OpenFlow compatible switch, and 1 controller. The hardware structure is depicted in Figure 3.7, where both hosts and the SDN controller exist as separate virtual machines on the same physical server. From Figure 3.7, the top row with the light green background represents 1 physical device: an HP 5900 Series switch (model JG336A)[48]. The HP switch has 48 1G/10GBase-T ports and is OpenFlow compatible (OpenFlow version 1.3 only). The bottom row with the light orange background represents 1 physical device: a SuperMicro SuperServer 8027R-TRF+. The SuperServer has an Intel Xeon processor E5-4600 v2 (12-Cores) with 8 individual 1000BASE-T RJ45 network interfaces. VMware ESXi communicates with the server hardware, and hosts the host virtual machines as well as the controller virtual machine. The host virtual machines have a Debian net install with basic python, ping, and tcpdump software.

Two controller VMs are made to host the controller software (representing the two parameters shown in Table 3.2). The Ryu controller offers a tutorial that recommends use of a saved VM image [49]. This image is used on the server to host Ryu. Ryu was selected because of its OpenFlow version 1.3 capability. Iris is another controller that is used because it can communicate using OpenFlow version 1.3 [21]. Iris is installed on an Ubuntu 12.04 Linux distribution virtual machine. Ryu and Iris alternate in suspended or active mode depending on which one needs to be running for a given collection of data.

As shown in Figure 3.7, 3 ports are connected from the HP switch to the server using 3 separate Ethernet cables, allowing a communication path that emulates a real network. The same communication occurs from end host to end host as is conducted within the MiniNet simulator.

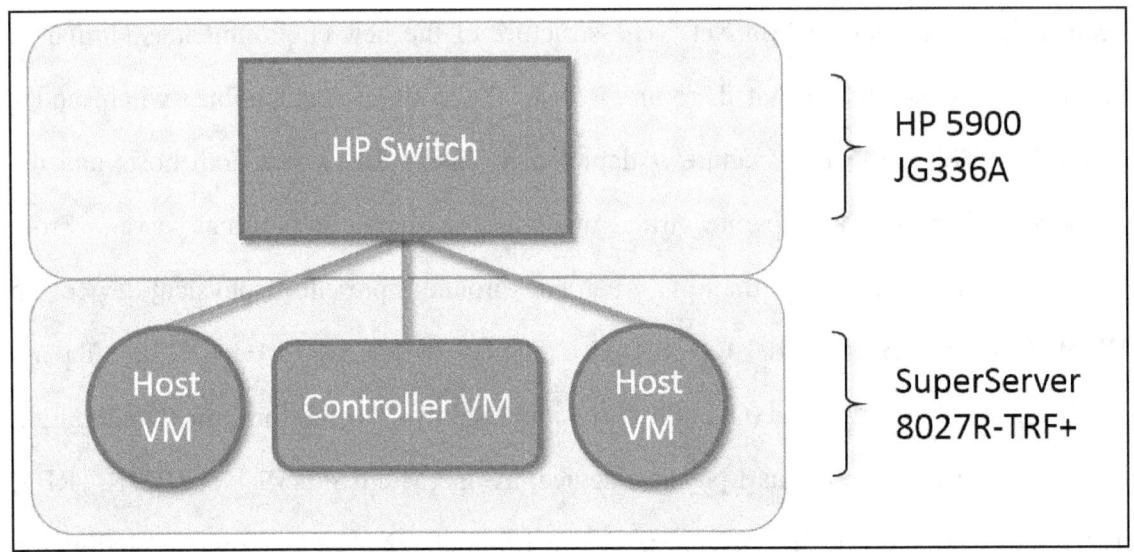

Figure 3.7: Hardware Simple Network

3.4.2 Experiment 2: Determine Flow Inactivity Timeout.

The goal of Experiment 2 is to determine whether a client within an SDN environment can accurately determine the inactivity timeout period for a particular flow. Determining the inactivity flow timeout period can assist in fingerprinting the SDN controller that is maintaining the SDN environment because it is another observable feature of the SDN controller. Experiment 2 uses the same latency technique described in Experiment 1 of determining whether a flow is still active within a switch.

3.4.2.1 Experiment 2 Background.

From Experiment 1, a client determines whether they are connected to an SDN or traditional network environment. After determining with 95% certainty that the network environment is an SDN environment, Experiment 2 determines the flow inactivity timeout period by systematically checking whether a flow is still installed after incremental periods

44

of time. Controllers have different default flow inactivity timeout periods, and consequently may be uniquely identifiable based on that inactivity timeout period. Additionally, if a malicious client were able to determine that the flow installed was never removed (or removed after a lengthy delay), that client may be able to target the specific type of flow as a means for denial of service by resource exhaustion of the network switch. The use of Experiment 2 for the purpose of a Denial of Service attack goes beyond the scope of this research and is suggested for future work.

3.4.2.2 Assumptions.

The assumptions for Experiment 2 are as follows:

1. The client has access to a port on a switch connected to the SDN environment.

2. ICMP echo packets are not blocked by the sending, receiving, or intermediary node.

3. Aggregate flows are not installed on the switch.

4. Aggregate flows are not used by the SDN controller.

5. The client is in an SDN environment.

6. The client knows the IP address of a live host that responds to ICMP echo requests, or has connected two devices that can successfully request and respond to ICMP echo packets.

7. The client has a reliable method for determining whether a flow is still active within an Openflow compliant switch (as demonstrated by Experiment 1).

3.4.2.3 Parameters.

Table 3.4 shows the parameters which define the treatment levels for both the simulated MiniNet test environment and the emulated test environment. The SDN controller parameter is none when the environment is a traditional environment because no controller is present. The hardware test environment has fewer SDN controllers because the hardware switch is compatible only with OpenFlow version 1.3, and the open-source SDN controllers listed in the simulated test environment use OpenFlow version 1.0. Table Table 3.5 shows the various parameters held constant to reduce the number of covariates. The response metric includes the difference between the predicted inactivity timeout value and the actual inactivity timeout value.

Table 3.4: Experiment 2 Parameters

Test Environment	Parameter	Setting
Simulated (MiniNet)	Environment	SDN
		Traditional
	SDN Controller	NOX
		POX
		Beacon
		Maestro
		Floodlight
		NodeFlow
		OpenDaylight
		None
Emulated (Hardware)	Environment	SDN
		Traditional
	SDN Controller	Ryu
		Iris
		None

46

Table 3.5: Experiment 2 Parameters Held Constant

Parameter	Value
Number of Clients	2
Number of Switches	1
Number of Controllers	1
Number of Hops between Host A and Host B	2

3.4.2.4 Hypothesis.

The expected output for Experiment 2 will show that the predicted value for the flow inactivity timeout period is within one second from the actual value of the flow inactivity timeout period.

3.4.2.5 Experiment Design.

Experiment 2 uses Mininet to simulate the SDN environment [25]. Within Mininet, a simple network consisting of two hosts attached to one switch with one controller (the control variable) is used. The simple network is the same network used in Experiment 1 and is depicted in Figure 3.6. Within Figure 3.6, The network switch, Hosts A and B are all Linux kernel version 3.8.0 emulated in a MiniNet virtual machine. The controller is hosted on a separate virtual machine running Ubuntu version 12.04 with Linux kernel version 3.8.0.

Host A represents the client that has no knowledge of the network environment. I deally, Host A first determines whether the environment is an SDN or traditional environment by utilizing the latency analysis methods discussed in Experiment 1. For the purposes of Experiment 2, the type of environment was known. Host A gathers a baseline mean and standard deviation of ICMP echo RTTs. Host A transmits 50 ICMP echo requests to Host B, and records the RTT for each request/response pair. The use of 50 is to gather a mean and standard deviation of RTTs, and the only requirement for this number is that it is greater than two (where the first RTT represents the flow installation latency, the second RTT represents the flow already installed, and subsequent RTTs allow calculations of a mean and standard deviation). Host A discards the first RTT collected, and calculates the mean and standard deviation of the remaining 49 RTTs. The calculated results are referred to as the baseline mean $(\overline{R_B})$ and baseline standard deviation (σ_B), and are shown in (3.1) and (3.2) respectively [50]. In both (3.1) and (3.2), R is the RTT of an ICMP echo request/response pair, and n is the number of RTTs recorded. The first RTT is discarded as it represents a flow-installation event, and would skew a baseline for when the flow is already installed.

$$\overline{R_B} = \frac{\sum (R)}{n} \tag{3.1}$$

$$\sigma_B = \sqrt{\frac{\sum (R - \overline{R_B})^2}{n-1}} \tag{3.2}$$

After recording a baseline mean and standard deviation, the inactivity timeout period is determined by first setting an estimate value of 2 seconds. Host A incurs a flow-installation event by sending an ICMP echo packet, then waits the estimate value of 2 seconds, then sends another ICMP echo packet, recording the RTT for this second packet. The recorded RTT is referred to as the new RTT (R_N). The Z_{Score} for the new RTT is

$$Z_{Score} = |R_N - \overline{R_B}|/\sigma_B \tag{3.3}$$

The meaning of the Z_{Score} response variable is shown in Table 3.6. If the calculated Z_{Score} is above a threshold value, then the observed RTT is significantly greater than the baseline mean, and likely represents a new flow installation event, which means the flow expired during the duration of the estimate value of two seconds. If the calculated Z_{Score} is below the threshold value, then the observed RTT is not significantly greater than the baseline mean, and indicates no new flow installation taking place, which means that the flow did not expire during the duration of the estimate value of two seconds. The threshold Z_{Score} value is experimentally set through trial and error to 50.

Table 3.6: Interpretation of Z_{Score} Values

Expression	Interpretation
$Z_{Score} > Z_{Score_{Threshold}}$	The estimate timeout is greater than the actual value
$Z_{Score} <= Z_{Score_{Threshold}}$	The estimate timeout is less than the actual value

Once the estimate timeout value is determined to be greater or less than the actual timeout value, the estimate is adjusted, and the process for calculating the Z_{Score} is repeated with the new estimate value. An example of adjusting the estimate value and repeating the Z_{Score} calculation is shown in Table 3.7.

Within Table 3.7, the initial minimum boundary for the estimate is zero (non-inclusive), and the maximum boundary for the estimate is infinite (∞). An initial estimate of 1 second is tested and found to be below the actual timeout value. The minimum boundary for the estimate is doubled from 1 to 2, and found to still be below the actual value. The estimate is doubled repeatedly until a Z_{Score} response indicates that the flow is absent, and thus the estimate is greater than the actual inactivity timeout value. When estimating 16 seconds, the flow is no longer present, indicating that the flow expired within 16 seconds, and therefore the flow inactivity timeout is less than 16 seconds. Because the flow timeout is less than 16 seconds, the estimate maximum boundary is decreased from infinity to 16 seconds. The estimate is adjusted from 16 seconds to halfway between the current value and the estimate minimum boundary of 8 seconds. The new estimate of 12 seconds is found to be greater than the actual inactivity timeout value, so the estimate maximum boundary is decreased from 16 seconds to 12 seconds and the estimate is again adjusted to halfway between the current value and the estimate minimum boundary of 8 seconds. The new estimate of 10 seconds is found to be below the actual value, so the estimate minimum boundary is increased from 8 to 10 and the estimate is adjusted to be halfway between the current value and the estimate maximum boundary of 12 seconds. The process continues until the difference between the estimate minimum boundary and estimate maximum boundary is less than or equal to the desired precision value.

It is important to note that this method of obtaining the inactivity timeout assumes there will not be a flow installation event due to a flow hard timeout. If the flow is found to be absent, but that absence is due to a hard timeout rather than an inactivity timeout, then the resulting inactivity timeout value may be incorrect. For Experiment 2, the hard timeout values were set to infinity to avoid this inconvenience. Given a hard timeout may exist on the SDN controller, a way to avoid an incorrect response is to wait N seconds between each inactivity test value, where N is the length of time determined to always cause a flow to expire (whether due to inactivity or a hard timeout). Waiting between each test case allows the flow to expire and thus resets the hard timeout counter.

Table 3.7: Discovering Inactivity Timeout (With Precision of 0.5 Seconds)

Estimate (seconds)	Result	Known Timeout Range (Seconds)
1	Flow is still installed	$1 < x < \infty$
2	Flow is still installed	$2 < x < \infty$
4	Flow is still installed	$4 < x < \infty$
8	Flow is still installed	$8 < x < \infty$
16	Flow is absent	$8s < x < 16$
12	Flow is absent	$8s < x < 12$
10	Flow is still installed	$10 < x < 12$
11	Flow is absent	$10s < x < 11$
10.5	Flow is absent, precision met	$10 < x < 10.5$

Experiment 2 utilizes the process shown in Table 3.7, and analyzes the resulting estimates compared with their true-values. Each controller is compiled with different flow inactivity timeout values. The set of flow inactivity timeout values includes 2, 25, 30, and 60 seconds. The upper-bound test value of 60 seconds was chosen because it is double the current largest known default inactivity timeout period (the Maestro SDN controller has a default timeout period of 30 seconds), and consequently provides a large range to interpolate for future SDN controller inactivity values.

Wireshark is used to ensure that the flow-modification packet sent from the SDN controller to the client contains the inactivity timeout field with the correctly set test value. The response data collected is then used to show a statistically significant correlation between the test inactivity timeout period and the response estimated inactivity timeout period. A statistically significant correlation combined with a difference of less than one second indicates that the method of estimating the inactivity timeout period is reliable for the timeout range.

The same experiment is conducted using the hardware environment previously described and depicted in Figure 3.7, where Hosts A and B perform the same actions as Host A and B respectively in the simulated Mininet environment, and the controller in Figure 3.7 performs the same action as the controller in the simulated Mininet environment. The hardware data collected is used to further support the conclusions obtained from the simulated experiments.

3.4.3   Experiment 3: Determine the Flow's Hard Timeout.

The goal of Experiment 3 is to determine whether a client within an SDN environment can accurately determine the hard timeout period for a particular flow. Determining a flow's hard timeout period can assist in fingerprinting the SDN controller that is maintaining the SDN environment as it is another observable feature within the feature table. Experiment

51

3 uses the same technique shown in Experiments 1 and 2 of determining whether a flow is still active within a switch.

3.4.3.1   Experiment 3 Background.

Experiment 3 is similar to Experiment 2, with the main difference being the value of interest. Experiment 3 shows that the flow hard timeout period can be reliably determined from the perspective of a client on an SDN network. The flow hard timeout period is determined by examining how long it takes for a flow to expire when constantly refreshed. A flow expires due to a hard timeout when the life of a flow exceeds the set hard timeout value. Unlike the inactivity period, the life of a flow cannot be reset by any packet-in event.

3.4.3.2   Assumptions.

The assumptions for Experiment 3 are as follows:

1. The client has access to a port on a switch connected to the SDN environment.

2. ICMP echo packets are not blocked by the sending, receiving, or intermediary node.

3. Aggregate flows are not installed on the switch.

4. Aggregate flows are not used by the SDN controller.

5. The client is in an SDN environment.

6. The client knows the IP address of a live host that responds to ICMP echo requests.

7. The client has a reliable method for determining whether a flow is still active within a switch (as demonstrated by Experiment 1).

3.4.3.3   Parameters.

Table 3.8 shows the parameters which define the treatment levels for both the simulated MiniNet test environment and the emulated test environment. The SDN controller parameter is none when the environment is a traditional environment because no controller is present. The hardware test environment has fewer SDN controllers because the hardware switch is compatible only with OpenFlow version 1.3, and the open-source SDN controllers listed in the simulated test environment use OpenFlow version 1.0. Table Table 3.9 shows the various parameters held constant to reduce the number of covariates. The response metric includes the difference between the predicted hard timeout value and the actual hard timeout value.

Table 3.8: Experiment 3 Parameters

Test Environment	Parameter	Setting
Simulated (MiniNet)	Environment	SDN
		Traditional
	SDN Controller	NOX
		POX
		Beacon
		Maestro
		Floodlight
		NodeFlow
		OpenDaylight
		None
Emulated (Hardware)	Environment	SDN
		Traditional
	SDN Controller	Ryu
		Iris
		None

Table 3.9: Experiment 3 Parameters Held Constant

Parameter	Value
Number of Clients	2
Number of Switches	1
Number of Controllers	1
Number of Hops between Host A and Host B	2

3.4.3.4 Hypothesis.

The expected output for Experiment 3 will show that the predicted value for the flow's hard timeout period is consistently within one second from the actual value of the flow's hard timeout period.

3.4.3.5 Experiment Design.

Experiment 3 uses Mininet to simulate the SDN environment [25]. Within Mininet, a simple network consisting of two hosts attached to one switch with one controller (the control variable) is used. The simple network is the same network used in Experiment 1 and 2, and is depicted in Figure 3.6. Within Figure 3.6, the network switch, Hosts A and B are all Linux kernel version 3.8.0 emulated in a MiniNet virtual machine. The controller is hosted on a separate virtual machine running Ubuntu version 12.04 with Linux kernel version 3.8.0.

Host A represents the client that has no knowledge of the network environment. Host A first determines whether the environment is an SDN or traditional environment by utilizing the latency analysis methods discussed in Experiment 1. As soon as the first ICMP echo request is sent to Host B from Experiment 1, Host A records a timestamp known as the first flow timestamp (f_0). After verifying that the environment is an SDN environment, Host A gathers a baseline mean and standard deviation of ICMP echo RTTs. Host A transmits 50 ICMP echo requests to Host B, and records the RTT for each request/response pair. The use of 50 is to gather a mean and standard deviation of RTTs, and the only requirement for this number is that it is greater than 2 (where the first RTT represents the flow installation latency, the second RTT represents the flow already installed, and subsequent RTTs allow calculations of a mean and standard deviation). Host A discards the first RTT collected, and calculates the mean and standard deviation of the remaining 49 RTTs. The calculated results are again referred to as the baseline mean ($\overline{R_B}$) and baseline standard deviation (σ_B), as was used during Experiment 2 as (3.1) and (3.2) respectively.

After recording a baseline mean and standard deviation, the hard timeout period is determined by sending an ICMP echo request every N seconds, where N represents the desired precision. N must be smaller than the inactivity timeout period determined from Experiment 2. If the inactivity timeout value has not been determined, any value that is known to be smaller than the inactivity timeout period results in the same outcome. For Experiment 3, N was set at 0.5 seconds. For each ICMP packet sent, the RTT for the ICMP packet is recorded. Using the same method in Experiment 2, if the RTT of the recent ICMP request compared to the previously recorded 49 RTTs results in a Z_{Score} (Equation 3.3) greater than a threshold Z_{Score} value, then a flow installation took place, indicating an expired flow entry. Given an ICMP packet is sent every 0.5 seconds, the flow expiration is not due to an inactivity timeout, but rather due to a hard timeout. When a hard timeout is observed, a new timestamp known as the flow expiration timestamp (f_E) is recorded. The hard timeout period (T_h) is calculated using

$$T_h = f_E - f_0 \qquad\qquad (3.4)$$

The threshold Z_{Score} value is experimentally set through trial and error to 50 as it was in Experiment 2. Table 3.10 shows an example process for determining a hard timeout period using a precision of 0.5 seconds (i.e., 0.5 second delay per ICMP echo request). Every 0.5 seconds, an ICMP RTT is recorded and used to determine whether the flow still resides in the switch. The process continues until a flow is determined to have expired, or an upper boundary is met. For Experiment 3, the upper boundary is six minutes (i.e., 360 seconds). If the flow does not expire after six minutes, the feature in the table is recorded as simply greater than six minutes. Six minutes is chosen as it is double the current longest known default hard timeout value in use by a controller (the Maestro controller has a default hard timeout value of 180 seconds), and consequently provides a large range to interpolate for future SDN controller hard timeout values. A value greater than six minutes is not used as an SDN controller may not have a hard timeout value, making the flow's hard timeout period effectively infinite, causing the experiment to take longer than necessary for testing values that may be infinite.

Table 3.10: Discovering Hard Timeout (With Precision of 0.5 Seconds)

Estimate (Seconds)	Result	Known Timeout Range (Seconds)
1.0	Flow is still installed	$1.0 < x < \infty$
1.5	Flow is still installed	$1.5 < x < \infty$
2.0	Flow is still installed	$2.0 < x < \infty$
2.5	Flow is still installed	$2.5 < x < \infty$
...	...	...
9.0	Flow is still installed	$9.0 < x < \infty$
9.5	Flow is still installed	$9.5 < x < \infty$
10.0	Flow is absent	$9.5s < x < 10.0$

Experiment 3 utilizes the process shown in Table 3.10, and analyzes the resulting estimates compared with their true values. Each controller is compiled with a flow hard timeout value from the set of test cases. The set of flow hard timeout values includes 10, 180, 200, and 360 seconds. The test scenarios are shown in Table 3.11. Ten seconds is

56

less than half of the shortest default hard timeout value for all SDN controllers used in Experiment 3 (i.e., Maestro has the shortest default hard timeout value of 30 seconds); 360 seconds is more than twice as long as the longest default hard timeout value for all SDN controllers used in Experiment 3.

Table 3.11: Experiment 3 Test Cases

Controller	Hard Timeout Test Cases (s)
POX	10, 180, 200, 360
Beacon	10, 180, 200, 360
Maestro	10, 180, 200, 360
Floodlight	10, 180, 200, 360
NodeFlow	10, 180, 200, 360

Wireshark is used to ensure the flow-modification packet sent from the SDN controller to the client contains the hard timeout field with the correctly set test value. The response data collected is then used to show a statistically significant correlation between the selected hard timeout period and the response estimated hard timeout period. A statistically significant correlation combined with a difference of less than one second indicates that the method of estimating the hard timeout period is reliable for selected timeout range.

The same experiment is conducted using the hardware environment previously described and depicted in Figure 3.7, where Hosts A and B perform the same actions as Host A and B respectively in the simulated Mininet environment, and the controller in Figure 3.7 performs the same action as the controller in the simulated Mininet environment. The collected hardware data is used to further support the conclusions obtained from the simulated experiments.

3.4.4 Experiment 4: Fingerprint SDN Controller.

The goal of Experiment 4 is to show that an unknown SDN controller can be identified by a client connected to the SDN environment. Experiment 4 uses the features collected in Experiment 2 and 3, along with other features of an SDN controller observable by a client of the SDN environment, to reliably distinguish different SDN controllers from each other.

3.4.4.1 Assumptions.

Considering Experiment 4 relies on all previous experiments in this thesis, all assumptions of Experiments 1-3 apply. For completeness, these assumptions are listed as follows:

1. The client has access to a port on a switch connected to the SDN environment.

2. ICMP echo packets are not blocked by the sending, receiving, or intermediary node.

3. Aggregate flows are not installed on the switch.

4. Aggregate flows are not used by the SDN controller. By default, all controllers used in this thesis do not use aggregate flows as aggregate flows require tailored administrative overhead, and are highly specialized for the network in which they are used.

5. The client is in an SDN environment. This assumption can be shown through Experiment 1.

6. The client knows the IP address of a live host that responds to ICMP echo requests. A live host responding to ICMP echo requests is required in order to determine packet RTTs.

7. The client has a reliable method for determining whether a flow is still active within a switch (as demonstrated by Experiment 1).

8. The SDN controller is listening on the default OpenFlow port of 6633.

9. The client knows the IP address of the SDN controller through the use of a successful TCP SYN scan for hosts with port 6633 active, and can communicate directly to the SDN controller.

Assumptions eight and nine were not included in Experiments 1-3 because those experiments involved communicating with another host connected to the network environment. Experiment 4 requires communicating directly with the SDN controller, making assumptions eight and nine necessary.

3.4.4.2 Parameters.

Table 3.12 shows the parameters which define the treatment levels. Table 3.13 shows the parameters held constant to reduce the possibility of covariates.

Table 3.12: Experiment 4 Parameters

Parameter	Setting
SDN Controller	NOX
	POX
	Beacon
	Maestro
	Floodlight
	NodeFlow
	OpenDaylight

3.4.4.3 Hypothesis.

The expected output for Experiment 4 will show that the controller behind an SDN environment can be determined from the perspective of a client connected to the SDN

59

Table 3.13: Experiment 4 Parameters Held Constant

Parameter	Value
Number of Clients	2
Number of Switches	1
Number of Controllers	1
Number of Hops between Host A and Host B	2

environment. Additionally, Experiment 4 will show that additional SDN controllers can be added to the feature table for querying which expands the fingerprint process to future controllers.

3.4.4.4 Experiment Design.

Similar to Experiments 1-3, Experiment 4 uses Mininet to simulate the SDN environment [25]. Within Mininet, a simple network consisting of two hosts attached to one switch with one controller (the control variable) is used. The simple network is the same network used in Experiments 1-3, and is depicted in Figure 3.6. Within Figure 3.6, the network switch, Hosts A and B are all Linux kernel version 3.8.0 emulated in a MiniNet virtual machine. The controller is hosted on a separate virtual machine running Ubuntu version 12.04 with Linux kernel version 3.8.0.

The controller for the SDN environment is chosen at random from a list of controllers and is started with its default installation configurations. The SDN controller listens on the default OpenFlow port (i.e., port 6633). After the OpenFlow compatible switch establishes a connection to the OpenFlow controller, the client at Host A in Figure 3.6 determines the IP address of the controller by performing a TCP SYN scan for the default OpenFlow port. Host A has a feature table that is populated with a list of SDN controllers. Host A's feature table is shown in Table 3.15. Within Table 3.15, the left-most column references the numerical entry found in the table of referenced keys shown in Table 3.14. The list of numbers shown in rows six and seven (corresponding to feature six and feature seven) of Table 3.15 represent different OpenFlow packet types, and are defined by Table 3.16. Each feature shown in Table 3.15 is described further in the following list.

1. The flow inactivity timeout period is the value shown observable by Experiment 2.

2. The flow hard timeout period is the value shown observable by Experiment 3.

3. After the TCP handshake is completed, but before the client sends any OpenFlow packets to the SDN controller, does the SDN controller send an OpenFlow Hello packet to the client? Given the OpenFlow Hello packet is a symmetric packet per the OpenFlow specification, the SDN controller implementation may wait for an initial OpenFlow Hello packet from the client before sending its own OpenFlow Hello Packet. This feature is a boolean value.

4. After the TCP handshake is completed, but before the client sends any OpenFlow packets to the SDN controller, does the SDN controller send any OpenFlow Echo requests? The OpenFlow specification uses OpenFlow Echo packets as keep-alive messages for the connection to remain persistent. These keep-alive messages occur after the Feature Response, yet SDN controller implementations may send them throughout the life of the connection, even before the symmetric OpenFlow Hello packets are received. This feature is a boolean value.

5. After the TCP handshake is completed, but before the client sends any OpenFlow packets to the SDN controller, does the SDN controller send an OpenFlow Feature request? The OpenFlow Feature request is initiated after the symmetric OpenFlow Hello exchange, yet SDN controller implementations may send the request before the exchange of OpenFlow Hello messages. This feature is a boolean value.

6. What packets are sent from the SDN controller after the client sends an OpenFlow Hello packet, but before any other packets from the client are sent? This feature in the table is stored as a list of numbers, with each number representing the packet type. The list of packet types is shown in Table 3.16 and is taken directly from the "type" field of the OpenFlow specification [12]. As an example shown in Table 3.15 row 6, after receiving an OpenFlow Hello packet, the NOX controller sends three OpenFlow packets (shown as 0, 5, 9): an OpenFlow Hello packet, followed by an OpenFlow Feature Request packet, followed by an OpenFlow Set Configuration packet.

7. What packets are sent from the SDN controller after the client sends an OpenFlow Hello packet, then a Feature Response packet, but before any other packets from the client are sent? This feature's value may depend on the contents of the feature response, thus a static feature response packet must be used that is the same for every SDN controller table entry.

Table 3.14: List of Keys

#	Key
1	Default Flow Inactivity-Timeout Period
2	Default Flow Hard-Timeout Period
3	Does the SDN controller initiate an OF Hello packet?
4	Does the SDN controller initiate an Echo Request?
5	Does the SDN controller initiate a Feature Request?
6	The set of packets sent by the SDN controller just after the Hello exchange
7	The set of packets sent by the SDN controller just after the Feature Response

62

Table 3.15: SDN Controller Features

#	NOX	POX	Beacon	Maestro	Floodlight	NodeFlow	OpenDaylight
1	5s	10s	5s	30s	5s	5s	∞
2	∞	30s	∞	180s	∞	∞	∞
3	False	True	True	True	True	False	True
4	False	False	True	False	False	False	False
5	False	False	False	False	False	False	False
6	0, 5, 9	5	5, 2, 2	5	5	0, 5	5, 5, 5
7	None	9, 14, 18	16, 2, 2	13, 13, 13	9, 18, 7	None	9, 7, 14

Each new feature added to the feature table is a key-value pair. The key is the set of actions that the client must perform, such as establishing a TCP connection followed by sending an OpenFlow Hello packet. The value is the expected response from the SDN controller, such as responding with a specific set of OpenFlow packets. Within Table 3.15, the key is the feature number (i.e., the left-most column) combined with the SDN controller type, while the value is the corresponding cell entry. The features shown in Table 3.15 were collected by downloading and installing each controller onto an Ubuntu 12.04 Linux Virtual Machine and connecting to them from the host machine using the python library.

Host A collects every feature of an unknown SDN controller and then compares that feature set to each known SDN controller in the table to determine a match. Features one and two are determined through methods described in Experiment 2 and 3 respectively. Features three through five are determined by establishing a TCP connection to the SDN controller and observing the response packets received. A python library of OpenFlow structures is created to communicate with SDN controllers directly. After establishing a TCP connection, Host A listens for 15 seconds and records any OpenFlow packets received by the SDN controller in the feature table (recorded as boolean features three, four, and five

Table 3.16: Partial List of OpenFlow Packet Types

Type	Descriptor
0	Hello
1	Error
2	Echo Request
5	Feature Request
7	Get Configuration
9	Set Configuration
10	Packet Input Notification
13	Packet Output
14	Flow Modification
15	Port Modification
16	Flow Statistics Request
18	Barrier Request

in Table 3.15). Fifteen seconds is chosen to allow a stopping point in the case where an SDN controller continuously sends OpenFlow packets. After collecting any desired feature, Host A begins collecting the next feature by killing the TCP connection and following the next key of actions. In the case of Experiment 4, Host A collects feature six by re-establishing a new TCP connection and then sending an OpenFlow Hello packet, observing any response. Host A records the response as feature six in the feature table. Host A again kills the TCP connection, and repeats the process of following a key of actions and recording the value responses in the feature table until every feature of the unknown SDN controller is collected.

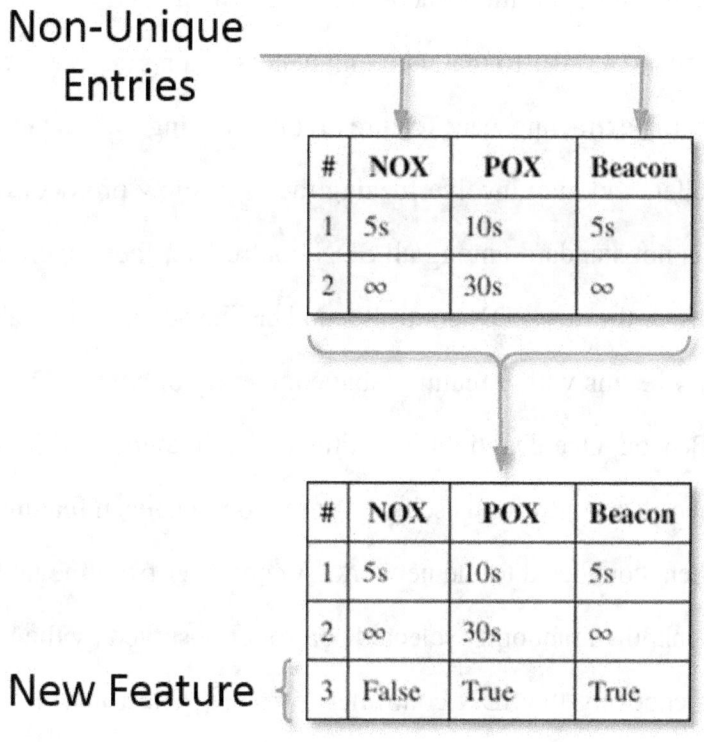

Figure 3.8: Ensuring Value Uniqueness

A value (i.e., set of features) uniquely identifies each SDN controller. A new SDN controller must have a unique value in order to be identified apart from another SDN controller. If two controllers have identical values, then adding the SDN controller to the table requires extracting a new feature that differentiates the new SDN controller from the similar SDN controller. Additionally, this new feature must be observable from the perspective of a client connected to the SDN environment, and must also be extracted from each SDN controller already existing in the table (avoiding null cells within the table). Figure 3.8 shows the abstract process for extracting a new feature in order to ensure that the set of features for each SDN controller is unique. The top of Figure 3.8 shows the Beacon SDN controller being added to the table, but its features are exactly the same as the NOX SDN controller, making them indistinguishable. As shown at the bottom of Figure 3.8, a third feature is extracted from the Beacon SDN controller (and consequently from every other SDN controller) in order to maintain uniqueness among table values.

The process for extracting new features requires using and experimenting with the new SDN controller, and may involve fuzzing the OpenFlow protocol. When a response is obtained that is not standard among all SDN controllers, the response is considered an identifying feature of the new SDN controller and can be added to the table of features.

Experiment 4 begins with a feature table containing only the SDN controllers NOX, POX, Maestro, Beacon, OpenDaylight, Floodlight and Nodeflow. With the simple network used in Experiment 1, a controller is selected at random and each feature of the network is observed by a client connected to the network. A controller from the table is then selected as matching the unknown randomly selected controller observed by the host. The selection process involves choosing the SDN controller from the table for which the most number of features is similar to the observed controller. The process is repeated sixteen times to determine the feasibility of correctly determining an unknown SDN controller. Sixteen is selected to allow each of the eight controllers a high chance of being selected at least once. The process continues if not every controller is selected through the first sixteen iterations.

IV. Results and Analysis

This chapter presents and analyzes the results from the experiments described in Chapter 3. Section 4.1 explains and analyzes the results of Experiment 1. Next, Section 4.2 and Section 4.3 describe the results of Experiments 2 and 3 in detail. Finally, Experiments 4 results are shown and analyzed in Section 4.4.

4.1 Experiment 1 Results and Analysis

The purpose of Experiment 1 is to determine whether it is possible to distinguish between a traditional network environment, and an SDN environment. Section 4.1.1 presents the data collected using the MiniNet simulation, while Section 4.1.2 presents the data collected using the hardware switch and controller.

4.1.1 Simulation Data.

Table 4.1 shows a subset of the ping RTTs captured. The first column is the controller, where a "Traditional" controller represents no controller present, and thus a traditional network environment without a controller. The second and third columns are the first and second ping RTTs in seconds respectively. The ratio column is the first ping RTT column divided by the second ping RTT column. The script that generates the data shown in Table 4.1 repeats the process 300 times for each controller, leading to the boxplots shown in Figure 4.1. Table 4.1 shows only the first iteration of each controller.

Table 4.2 lists the summary of the 300 RTT ratios collected and presented in Figure 4.1. The summary data in Table 4.2 includes outliers, and shows the ratio's minimum, maximum, mean ($\overline{x}_{ratio}$), and standard deviation (δ_{ratio}) values. From the summary data, the traditional network environment has a mean RTT ratio near one, which indicates that the first ICMP echo RTT and the second ICMP echo RTT are similar in value. The mean RTT ratio of other SDN controllers is greater than one, with larger standard

67

deviations, indicative of the delay incurred by a packet-in and packet-out event for flow modifications.

Table 4.1: Experiment 1 Sample Data

Controller	First Ping RTT (s)	Second Ping RTT (s)	Ratio
NOX	0.002452	0.000051	48.07844
Maestro	0.000418	0.000017	24.58824
Beacon	0.005686	0.000040	142.1500
POX	0.005661	0.000041	138.0732
Floodlight	0.003026	0.000030	100.8667
NodeFlow	0.026531	0.000041	647.0976
Traditional	0.001122	0.000557	2.014363

In order to distinguish between a traditional network environment and an SDN environment, it must be shown that the RTT ratio shown in Table 4.1 is significantly less in a traditional network environment than in any SDN environment. Before determining a statistical test to answer whether the RTT is significantly less in a traditional network, it is important to view a graphical summary of data to validate assumptions for each available statistical test. Figure 4.1 contains a box plot for the traditional network environment data, as well as an individual box plot for each controller used in Experiment 1. Figure 4.1 shows that each box plot has a very different distribution of RTT ratios.

4.1.1.1 Analysis.

It is difficult to determine whether the traditional network RTT ratios are significantly less than the SDN RTT ratios from Figure 4.1 because the range of RTT ratios corresponding to the NodeFlow controller increases the y-axis scale to the thousands,

Table 4.2: Experiment 1 Summary Data

Controller	ratio$_{min}$	ratio$_{max}$	$\bar{x}_{ratio}$	δ_{ratio}
NOX	26.85714	158.077	68.65388	20.86483
Maestro	0.009978617	2205.864	38.04614	141.9836
Beacon	1.534444	534.122	140.4314	67.81956
POX	2.928572	1452.731	187.9344	364.5844
Floodlight	76.02439	458.500	146.878	49.98321
NodeFlow	301.2093	3878.609	1352.519	484.2968
Traditional	0.2880355	2.790476	0.9755106	0.3718565

shrinking the traditional environment boxplot. Comparing the RTT ratios requires a statistical test.

The analysis of variance F-test allows comparing data from several means. With the F-test, data is compared to answer whether there exists differences between any means. Ideally, the F-test is applied across all groups of data (traditional, NOX, Maestro, etc.), and shows that a difference exists. The F-test is applied across only the groups of controllers (all groups of data except data from the traditional network environment), and used to show that no difference exists. If a difference is observed with the traditional data set included, and then no difference observed when the traditional data set is excluded, a conclusion can be made declaring that the traditional network RTT ratios deviate from the SDN RTT ratios.

While the F-test is not sensitive to the normality of data (so long as the sample size is sufficiently large), there is an assumption of uniform standard deviations between groups. Figure 4.1 presents highly non-uniform standard deviations of the RTT ratios for each controller. The appearance is further verified in Table 4.2, where δ_{ratio} represents the sample standard deviation of the corresponding controller RTT ratios. Because of the non-uniform standard deviation, an assumption of the F-test is violated, and the analysis is reduced to comparing each group independently.

While the F-test is not sensitive to the normality of data (so long as the sample size is sufficiently large), there is an assumption of uniform standard deviations between groups. Figure 4.1 presents highly non-uniform standard deviations of the RTT ratios for each controller. The appearance is further verified in Table 4.2, where δ_{ratio} represents the sample standard deviation of the corresponding controller RTT ratios. Because of the non-uniform standard deviation, an assumption of the F-test is violated, and the analysis is reduced to comparing each group independently.

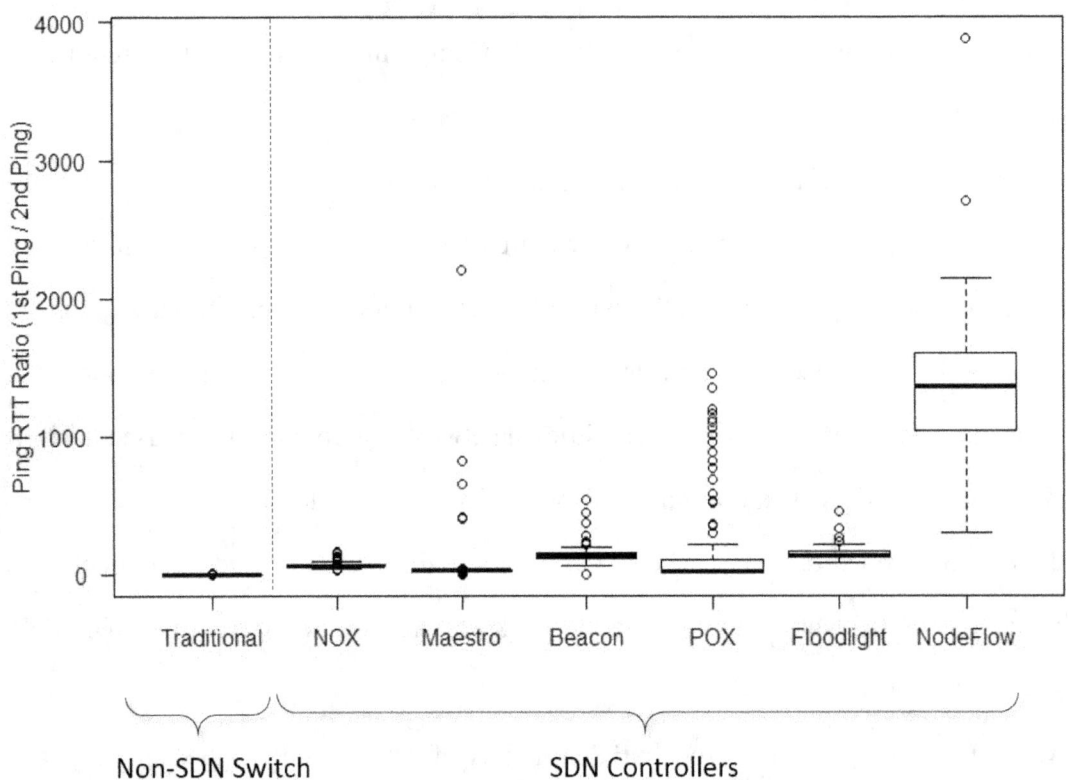

Figure 4.1: Ping Ratios with Simulated SDN Controllers

The Student's t-test is used to quantify the difference between the traditional network environment's RTT ratios and the SDN environment's RTT ratio [50]. With the t-test, the sample mean and standard deviation of a population of interest are recorded. A new sample of a different population is taken, where the new mean is compared with the original mean. The difference between the two means, in units of standard deviations of the former sample, represents the parameter used in the t-test. The parameter is the Z_{score} and is

$$Z_{Score} = \frac{\overline{x}_{SDN} - \overline{x}_{traditional}}{s_{traditional}} \tag{4.1}$$

where $\overline{x}_{SDN}$ represents the mean RTT ratio of the selected SDN controller, $\overline{x}_{traditional}$ represents the mean RTT ratio of the traditional environment, and $s_{traditional}$ represents the sample standard deviation of the RTT ratios for the traditional environment.

In terms of Experiment 1, the null hypothesis states that the traditional network environment RTT ratios are the same as the SDN environment RTT ratios. When performing the Student's t-test, the p-value represents the probability of collecting a sample with the resulting mean through random chance alone. A p-value less than 0.05 is sufficient to reject the null hypothesis. Table 4.3 shows the resulting p-values from applying the Student's t-test to each SDN environment.

Table 4.3: Experiment 1 T-Test Results

Controller	Z_{Score}	P-Value
NOX	182.001	0
Maestro	99.691	0
Beacon	375.026	0
POX	502.772	0
Floodlight	392.026	0
NodeFlow	3634.583	0

71

From Table 4.3, every controller results in a ping RTT ratio greater than 90 standard deviations (i.e., a Z_{Score} greater than 90) away from the mean value of the traditional environment ping RTT ratio. The Z_{Score} values applied to the Student's t distribution result in a p-value very near zero, providing strong evidence to reject the null hypothesis.

4.1.1.2   Other Observations.

An interesting observation to note involves the Maestro controller data. In Table 4.2, the Maestro controller has a $ratio_{min}$ value of 0.00998, which is less than the Traditional environment's $ratio_{max}$ value of 2.79048. The fact that an SDN controller's RTT ratio value at one point is less than the traditional environment's $ratio_{max}$ value indicates the potential for false positives. For instance, if a host is attempting to identify whether the network is a traditional or an SDN environment, collection of an RTT ratio of 0.00998 leads to the false conclusion that the Maestro-controlled SDN network is a traditional network. The idea of false positives is depicted in Figure 4.2, which displays hypothetical data. Within Figure 4.2, the range of minimum and maximum RTT ratios are displayed with an arrow between both values. The $ratio_{max}$ of the traditional network environment creates the threshold where ideally all SDN environment RTT ratios do not fall below. The potential for false positives exists when the traditional $ratio_{max}$ value is greater than the SDN $ratio_{min}$ value, as is the case for SDN 3. The probability of false positives increases when the former potential exists, and the difference (depicted as δ in Figure 4.2) between the traditional $ratio_{max}$ value and any SDN $ratio_{min}$ value increases.

While the $ratio_{min}$ value for the Maestro controller is an outlier, an analysis is performed to determine the robustness of the method for distinguishing SDN environments from traditional environments. The analysis is the percentage of SDN RTT values that exist below the traditional environment's maximum RTT ratio value. These percentages are shown in Table 4.4. From Table 4.4, only controllers Maestro and Beacon exhibit RTT values that fall below the traditional environment's RTT threshold. Between Maestro

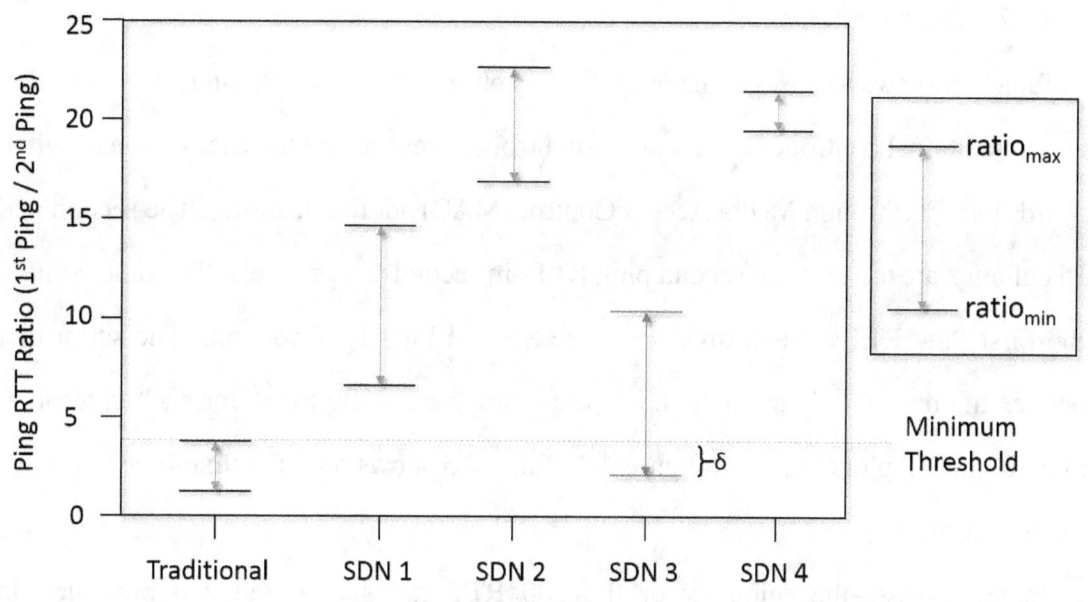

Figure 4.2: Example Ping Ratios

and Beacon, less than 2 percent of their RTT values fall below the traditional environment's RTT threshold, providing a strong indication that the potential for false positives is minimal.

Table 4.4: Experiment 1 RTT Values Below Maximum Traditional RTT Value

Controller	% Below Threshold
NOX	0.00%
Maestro	1.67%
Beacon	1.01%
POX	0.00%
Floodlight	0.00%
NodeFlow	0.00%

The data from Experiment 1 strongly indicate that it is possible for a host connected to a network to distinguish between an SDN and traditional network environments.

4.1.2 Hardware Data.

Table 4.5 shows a subset of the ping RTTs captured. The first column is the controller, where a traditional controller represents no controller present, and thus a traditional switch forwards packets through Media Access Control (MAC) address learning. The second and third columns are the first and second ping RTTs in seconds respectively. The ratio column is the First Ping RTT column divided by the Second Ping RTT column. The script that generates the data shown in Table 4.5 repeats the process 300 times for each controller, leading to the boxplots shown in Figure 4.3. Table 4.5 shows only the first three iterations of each controller.

Table 4.6 lists the summary of the 300 RTT ratios collected and presented in Figure 4.3. The summary data in Table 4.6 includes outliers, and shows the ratio's minimum, maximum, mean, and standard deviation values. From the summary data, the traditional network environment has a mean RTT ratio near one, which indicates that the first ICMP echo RTT and the second ICMP echo RTT are similar in value. The mean RTT ratio of other SDN controllers is greater than one, with larger standard deviations, indicative of the delay incurred by a packet-in and packet-out event for flow modifications.

4.1.2.1 Analysis.

In order to distinguish between a traditional network environment and an SDN environment, it must be shown that the RTT ratio shown in Table 4.5 is significantly less in a traditional network environment than in any SDN environment. Figure 4.3 contains a box plot for the traditional network environment data, as well as an individual box plot for each controller used in Experiment 1. Figure 4.3 shows that each box plot has a very different distribution of RTT ratios, and also indicates that the SDN environment ping RTT ratios are higher in value.

74

Table 4.5: Experiment 1 Sample Hardware Data

Controller	First Ping RTT (s)	Second Ping RTT (s)	Ratio
Ryu	0.015655	0.000596	26.2667786
Ryu	0.014527	0.000522	27.8295018
Ryu	0.019545	0.000493	39.6450305
Iris	0.013426	0.000408	32.9068628
Iris	0.014358	0.000506	28.3754939
Iris	0.016179	0.000291	55.59793741
Traditional	0.000614	0.000535	1.1476635
Traditional	0.000564	0.000465	1.2129032
Traditional	0.000353	0.000570	0.6192982

Table 4.6: Experiment 1 Hardware Summary Data

Controller	$ratio_{min}$	$ratio_{max}$	$\bar{x}_{ratio}$	s_{ratio}
Ryu	8.268838	69.53261	34.55759	10.30061
Iris	10.57498	92.26437	25.19955	10.57498
Traditional	0.6192982	1.884244	1.100641	0.2161701

The Student's t-test is used to quantify how much higher the SDN ping RTT ratios are when compared to the traditional environment's ping RTT ratios [50]. With the t-test, the sample mean and standard deviation of a population of interest are recorded. A new sample of a different population is taken, where the new mean is compared with the original mean. The difference between the two means, in units of standard deviations of the former sample, represents the parameter used in the t-test. The parameter is the Z_{score}, and is shown under Section 4.1.1.1 in (4.1), where $\bar{x}_{SDN}$ represents the mean RTT ratio of the selected

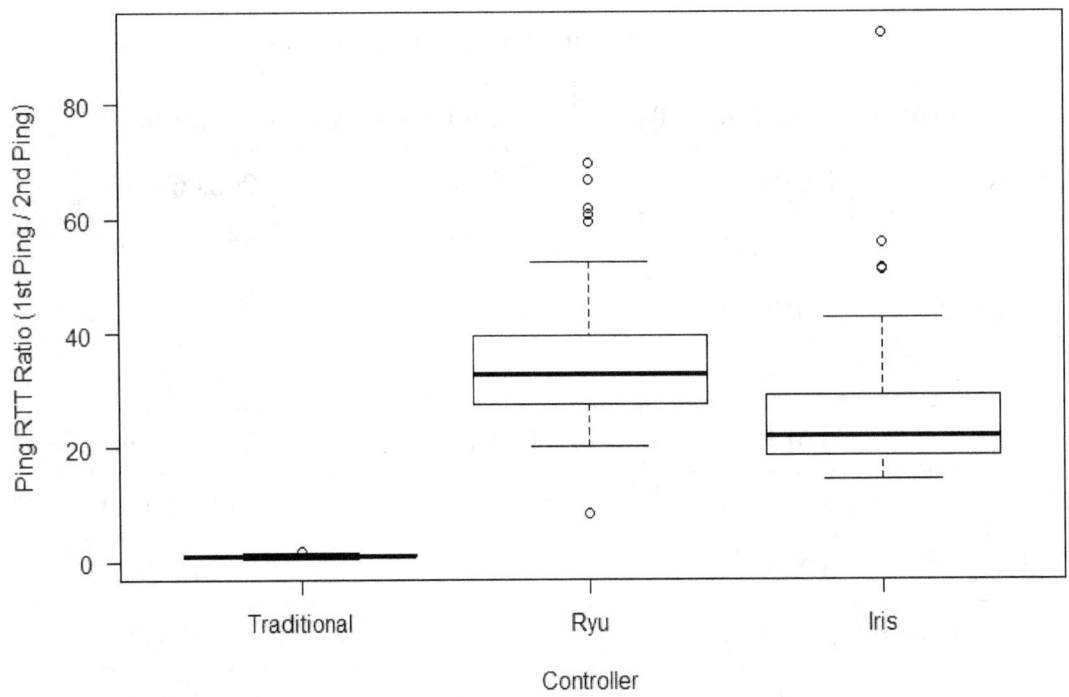

Figure 4.3: Ping Ratios with Emulated SDN Controllers

SDN controller, $\bar{x}_{traditional}$ represents the mean RTT ratio of the traditional environment, and $s_{traditional}$ represents the sample standard deviation of the RTT ratios for the traditional environment.

In terms of Experiment 1 hardware data, the null hypothesis states that the traditional network environment RTT ratios are the same as the SDN environment RTT ratios. When performing the Student's t-test, the p-value represents the probability of collecting a sample with the resulting mean through random chance alone. A p-value less than 0.05 is sufficient to reject the null hypothesis. Table 4.7 shows the resulting p-values from applying the Student's t-test to each SDN environment.

From Table 4.7, both controllers result in a ping RTT ratio greater than 100 standard deviations (i.e., a Z_{Score} greater than 100) away from the mean value of the traditional

Table 4.7: Experiment 1 T-Test Results

Controller	Z_{Score}	P-Value
Ryu	154.7714	0
Iris	111.4812	0

environment ping RTT ratio. The Z_{Score} values applied to the Student's t distribution result in a p-value very near zero, providing strong evidence to reject the null hypothesis.

4.1.2.2 Other Observations.

When comparing the results of the simulated data with the results of Experiment 1 applied with the HP switch, both indicate a reliable capability in distinguishing SDN environments from traditional environments. It is also interesting to note that the minimum RTT ratio observed for both Ryu and IRIS controllers does not fall below the maximum RTT ratio for the traditional environment, providing supporting evidence that there is a low probability for a false positive (i.e., low probability that an SDN environment is misidentified as a traditional environment).

4.2 Experiment 2 Results and Analysis

The purpose of Experiment 2 is to determine whether a host can successfully determine the controller's inactivity timeout value of the SDN environment. Section 4.2.1 presents the data collected using the MiniNet simulation, while Section 4.2.2 presents the data collected using the hardware switch and controller.

4.2.1 Simulated Data.

Table 4.8 shows a subset of the collected data. The first column specifies the controller used in the SDN environment. The second column is the inactivity timeout value in seconds that was programmed into the controller. The third column displays the response variable: the predicted inactivity timeout value in seconds obtained by the procedure outlined in

Chapter 3. The fourth column shows the difference between the predicted inactivity timeout and the actual inactivity timeout values recorded as an absolute value. The difference in values shows how accurate the prediction method was in determining the actual inactivity timeout value, where a smaller difference represents a better estimation. The fifth column is the elapsed runtime in seconds for obtaining the predicted inactivity timeout value. The elapsed runtime data is recorded for future research attempting to create a faster process in determining the SDN controller's timeout value.

Table 4.8: Experiment 2 Sample Data

Controller	Actual Timeout (s)	Predicted Timeout (s)	$\lvert\Delta\rvert$(s)	Runtime (s)
POX	2.00	2.6250	0.625	14.43342
POX	25.00	25.3750	0.3750	216.14472
Beacon	2.00	2.6250	0.625	14.33750
Beacon	25.00	25.6250	0.625	216.88608
Maestro	2.00	2.6250	0.625	14.30233
Maestro	25.00	25.6250	0.625	216.52465

Table 4.9 shows a summary of the collected data for Experiment 2. The controller column represents the specific controller in use while the data was collected. Under the controller column, "combined" represents the summary data of all controllers, while "combined (no outliers)" includes all data minus the three outliers shown in the histogram (Figure 4.4). The Δ_{min} column shows the minimum difference observed between the predicted timeout and actual timeout values, while the Δ_{max} column shows the maximum difference observed between the predicted timeout and actual timeout values. The $\overline{\Delta}$ column shows the mean difference observed between the predicted timeout and actual timeout values, and the s_Δ column shows the sample standard deviation.

Table 4.9: Experiment 2 Summary Data

Controller	$\Delta_{min}(s)$	$\Delta_{max}(s)$	$\overline{\Delta}(s)$	s_Δ
POX	0.125	48.3745	3.649975	11.20959
Beacon	0.125	17.125	1.325	3.725446
Maestro	0.125	0.625	0.4625	0.1677051
Combined	0.125	48.3745	1.812492	6.84007
Combined (no outliers)	0.125	1.125	0.44956	0.188749

Figure 4.4 displays a histogram showing the distribution of values for the timeout delta. Given three values appear as outliers in Figure 4.4, Figure 4.5 was generated to view the distribution of timeout delta values among the 57 values between zero and ten. The affect of the outliers as well as attribution to the cause of the outliers is discussed in Section 4.2.1.1.

4.2.1.1 Analysis.

Determining a relationship between two samples involves calculating Pearson's r correlation coefficient, which is a value between -1 and 1 inclusive and describes the degree of linear association between two variables [50]. Values close to zero indicate no correlation, and thus no observable relationship between the two variables, while values near positive or negative one show a stronger correlation between both variables, and thus a strong observable relationship between both variables. Pearson's r correlation coefficient is

$$r_{XY} = \frac{1}{n-1} \sum_{i=1}^{n} \left(\frac{X_i - \overline{X}}{s_X} \right) \left(\frac{Y_i - \overline{Y}}{s_Y} \right) \tag{4.2}$$

Applying (4.2) to the data in Experiment 2, X represents the sample of actual timeout values that were programmed into the controller, while Y represents the response variable

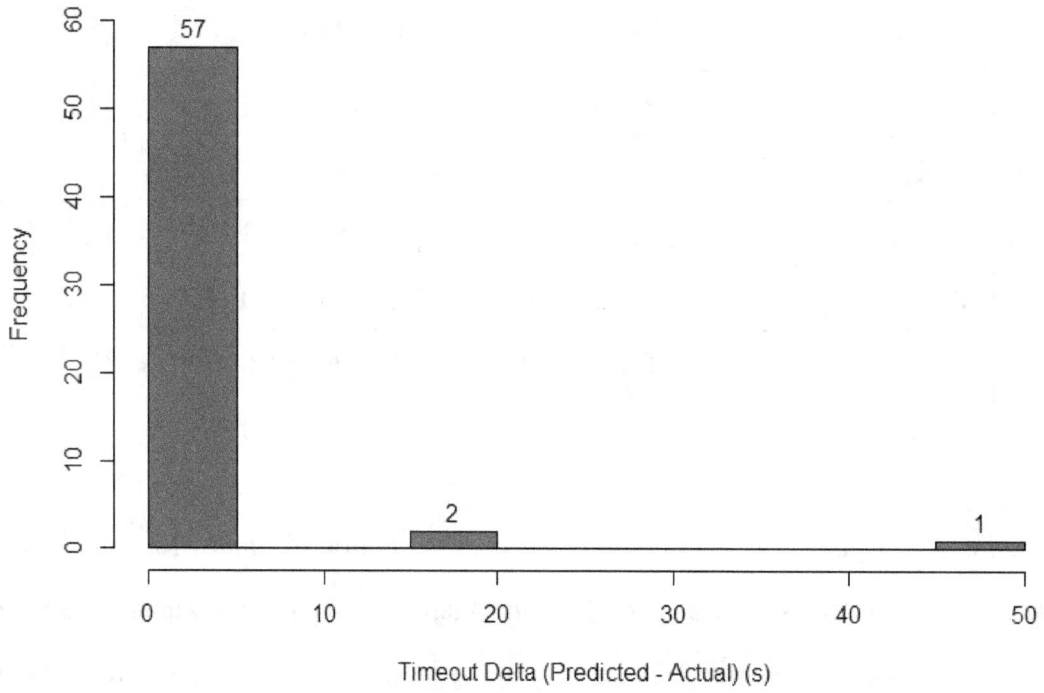

Figure 4.4: Inactivity Timeout Delta Histogram (Outliers Included)

sample: the predicted timeout values. The value n represents the sample size of both X and Y combined, while s represents the sample standard deviation.

Pearson's r correlation coefficient for both sample sets, with outliers included, is 0.9434948, which indicates a strong positive correlation between the predicted timeout value and the actual timeout value. The existence of a strong correlation, along with a mean delta of 1.8125 seconds, indicates that the method of predicting the actual timeout value has a precision of less than 2 seconds. The standard deviation of both sample sets is 6.84, which indicates a lack of reliability in determining the inactivity timeout with a desired precision of less than 1 second.

Outliers heavily affect the mean, standard deviation, and correlation coefficient. With outliers (i.e., the three delta values of 17.125, 17.625, and 48.3745) omitted, Pearson's r correlation coefficient increases to 0.9878111, with a mean delta of 0.4495614 and a

80

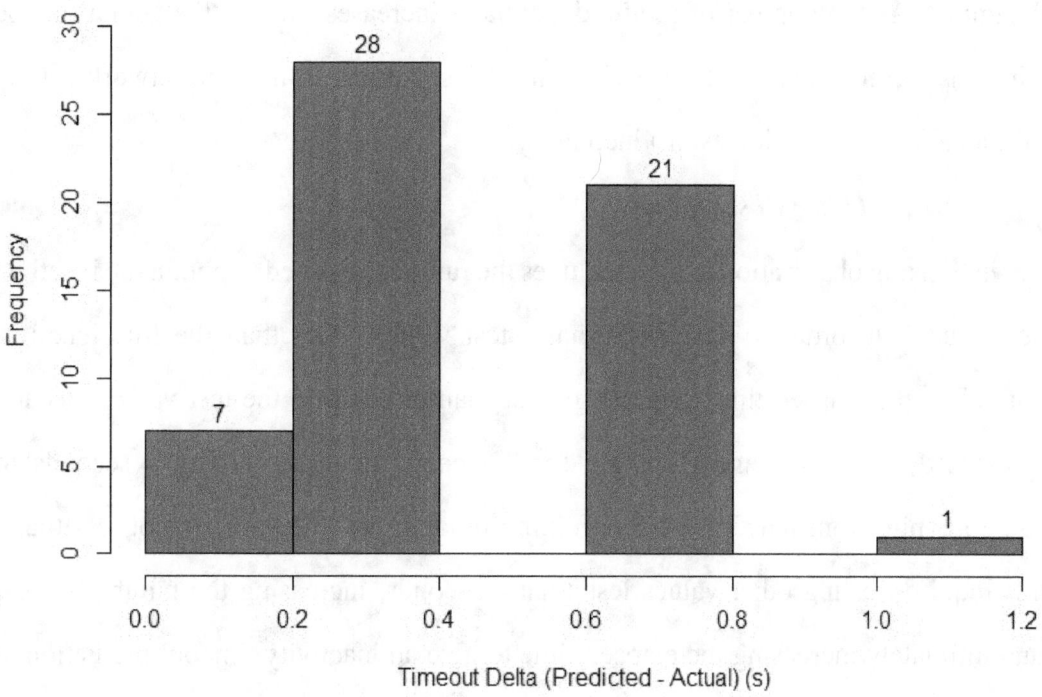

Figure 4.5: Inactivity Timeout Delta Histogram (Outliers Excluded)

standard deviation of 0.18877, indicating strong precision (i.e., a precision of less than 1 second) in predicting the actual inactivity timeout value for a controller. A 95% confidence interval of the mean delta value with outliers omitted is between 0.39947 and 0.49965.

Outliers can be attributed to network latency spikes. Considering that the method for determining the inactivity timeout relies on taking an initial sample of RTT values for an ICMP packet, then detecting any spike in a new RTT value, a spike due to normal networking lag rather than the installation of a new flow would cause this method to generate a false predicted timeout value. Depending on when the spike occurred, the resulting inaccurate response may be anywhere between zero and the actual inactivity timeout value. Referring back to Table 3.7, the estimates column indicates when a new ICMP packet is transmitted and an RTT is recorded, which means that at each of those times (i.e., 1 second, 2 seconds, 4 seconds, etc.) the RTT is tested for deviation from the initial sample. As the number of required estimates increases (i.e., as the actual timeout value increases), the chances of observing an RTT spike due to normal network activity (rather than a flow installation event) increases.

4.2.1.2   Other Observations.

An interesting observation to note includes the runtime required to obtain the inactivity timeout values. In order to determine that a test value is less than the true inactivity timeout value, the elapsed time must be greater than or equal to the test value. Because of the elapsed time requirement, testing for 8 seconds requires waiting 8 seconds for the flow to expire from inactivity. Determining reliably that the value is not less than 8 seconds requires testing other values less than 8 seconds, increasing the number of tests run, and ultimately increasing the elapsed time to give an inactivity timeout prediction. A hypothetical example of determining the value of 2.625 is shown in Table 4.10, which has the same layout as Table 3.7. Each value in the estimate column requires a runtime equal to the value shown, so to determine the final estimated value of 2.625 seconds, the sum of all estimates is taken (i.e., $2+4+3+2.5+2.75$), resulting in a runtime of 14.25 seconds. From data shown in Table 4.8, for when the predicted value is 2.6250 seconds, the runtime is between 14.3 and 14.5 seconds, which is not far from the calculated elapsed time of 14.25 seconds.

Table 4.10: Discovering Inactivity Timeout of 2.625 (With Precision of 0.25 Seconds)

Estimate (Seconds)	Result	Known Timeout Range (Seconds)
2	Flow is still installed	$2 < x < \infty$
4	Flow is absent	$2 < x < 4$
3	Flow is absent	$2 < x < 3$
2.5	Flow is still installed	$2.5 < x < 3$
2.75	Flow is absent, precision met	$2.5 < x < 2.75$

4.2.2   Hardware Data.

Table 4.11 shows a subset of the collected data with the hardware environment. The first column specifies the controller used in the SDN environment. The second column is the inactivity timeout value in seconds that was programmed into the controller. The third column displays the response variable: the predicted inactivity timeout value in seconds obtained by the procedure outlined in Chapter 3. The fourth column shows the difference between the predicted inactivity timeout and the actual inactivity timeout values recorded as an absolute value. The difference in values shows how accurate the prediction method was in determining the actual inactivity timeout value, where a smaller difference represents a better estimation. The fifth column is the elapsed runtime in seconds for obtaining the predicted inactivity timeout value.

Table 4.11: Experiment 2 Sample Hardware Data

Controller	Actual Timeout (s)	Predicted Timeout (s)	$|\Delta|$(s)	Runtime (s)
Ryu	10.00	11.125	1.125	85.96292
Ryu	30.00	54.625	24.625	500.965
Iris	10.00	10.375	0.375	83.981581
Iris	30.00	32.125	2.125	382.384175

Table 4.12 shows a summary of the collected data for Experiment 2. The controller column represents the specific controller in use while the data was collected. Under column controller, "combined" represents all the data combined. Δ_{min} column shows the minimum difference observed between the predicted timeout and actual timeout values, while the Δ_{max} column shows the maximum difference observed between the predicted timeout and actual timeout values.

Table 4.12: Experiment 2 Summary Hardware Data

Controller	$\Delta_{min}(s)$	$\Delta_{max}(s)$	$\overline{\Delta}(s)$	s_Δ
Ryu	0.625	49.375	20.66667	17.85885
Iris	0.125	7.125	2.927083	2.592819
Combined	0.125	49.375	11.79688	15.48267

Figure 4.6 shows a histogram presenting the distribution of values for the inactivity timeout delta (i.e., the difference between the actual and predicted inactivity timeouts). The frequency axis represents the number of occurrences in which a particular value was observed, while the timeout delta axis provides the range of inactivity timeout values in seconds. Each bar represents the number of values existing between a range of 5 values with the lower boundary inclusive and the upper boundary exclusive.

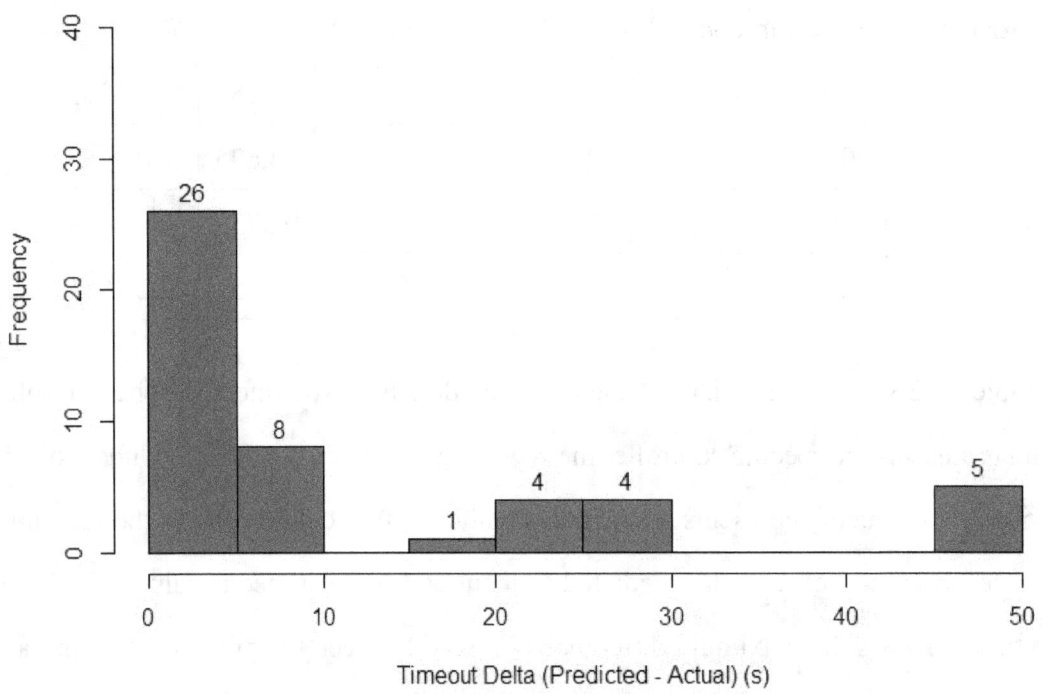

Figure 4.6: Inactivity Timeout Delta with HP Switch Histogram

84

4.2.2.1 Analysis.

Figure 4.6 shows several delta values existing greater than the desired precision of one second. From this chart alone the data proves inconsistent and therefore shows that the inactivity timeout prediction method is not reliable. Table 4.13 shows a closer look at a particular set of data, which is consistently incorrect. Within Table 4.13, the controller column describes the controller from which the data is obtained. The actual timeout column is the inactivity timeout value programmed into the controller, and the predicted timeout column is the response variable: the predicted inactivity timeout value in seconds obtained by the method outlined in Chapter 3.

Table 4.13: Experiment 2 Hardware Data Subset

Controller	Actual Timeout (s)	Predicted Timeout (s)	Δ(s)
Ryu	10	11.125	1.125
Ryu	10	14.875	4.875
Ryu	10	15.875	5.875
Ryu	10	15.875	5.875
Ryu	10	15.875	5.875
Ryu	60	109.375	49.375
Ryu	60	109.375	49.375
Ryu	60	109.375	49.375
Ryu	60	109.375	49.375
Ryu	60	109.375	49.375

The actual inactivity timeout is not only different than the predicted inactivity timeout by greater than the desired precision of one second, but it is also consistently greater. For every attempt at predicting the true inactivity timeout value when the actual inactivity value was set to 60, the response was a consistent 109.375 seconds, causing the predicted value to be off by more than 49 seconds. Additionally, as the actual inactivity timeout value increases, the size of the error increases.

The inactivity timeout value set for the HP switch is verified to ensure that the controller is transmitting the correct inactivity timeout value for a new flow installation. Polling the switch for its flow table results in flow information confirming the inactivity timeout value expected. The fact that the HP switch responds with the correctly set inactivity timeout value suggests that the HP switch does not stringently enforce its timeout policy. An inactivity timeout value of 60 seconds takes more than 60 seconds of inactivity for the HP switch to actually remove the flow.

4.2.2.2   Other Observations.

An interesting observation is the different response variables obtained when different controllers have the same inactivity timeout value programmed. Table 4.14 shows both Ryu and Iris controllers with an actual timeout value set to 60 seconds. When a flow installation from the Ryu controller tells the HP switch to set an inactivity timeout of 60 seconds, the HP switch deletes the flow consistently after more than 100 seconds pass. When a flow installation from the Iris controller tells the HP switch to set an inactivity timeout of 60 seconds, the HP switch deletes the flow consistently after only 64 seconds pass. The reason for the disparity may exist in the flow installation request packet sent by the controller, as the HP switch actions depend on the requests from the controller.

Figure 4.7 and Figure 4.8 both show the output when the HP switch is sent the command "display openflow instance 1 flow-table" with the exception that for Figure 4.7, the attached controller is the Ryu controller, and with Figure 4.8 the attached controller is the Iris controller. Both command responses have been clipped short to only include the pertinent flow entries. The flow entry that is not shown handles messages bound for the controller (i.e., packet-in messages generated from packets that the switch does not know

86

Table 4.14: Experiment 2 Hardware Data Subset

Controller	Actual Timeout (s)	Predicted Timeout (s)	Δ(s)
Ryu	60	109.375	49.375
Ryu	60	109.375	49.375
Ryu	60	109.375	49.375
Ryu	60	109.375	49.375
Ryu	60	108.375	48.375
Iris	60	64.125	4.125
Iris	60	64.125	4.125
Iris	60	64.125	4.125
Iris	60	64.125	4.125
Iris	60	64.125	4.125

how to handle). Disparities between the two outputs are shown boxed in red. One disparity (aside from the match information) is the flags set by both controllers. The Ryu controller sets no flags for the flow installation, while the Iris controller sets a "flow_send_rem" flag. The "flow_send_rem" flag dictates that a flowremoved message is transmitted to the controller when a flow entry is removed or expires [51]. Another disparity between Figure 4.7 and Figure 4.8 is the priority, which is documented by HP as "Priority of the flow entry. The larger the value, the higher the priority" [51].

The HP switch's implementation of the flow entry's inactivity timeout may differ based on the flag or priority or some other difference. Both flows have an inactivity time (i.e., the time labelled as an "idle time" in Figure 4.7 and Figure 4.8) of 60 seconds, yet the time with which they are deleted from the flow table differs (as shown in Table 4.14). Determining the specific cause of the difference is difficult as implementation details beyond the HP switch documentation is not provided.

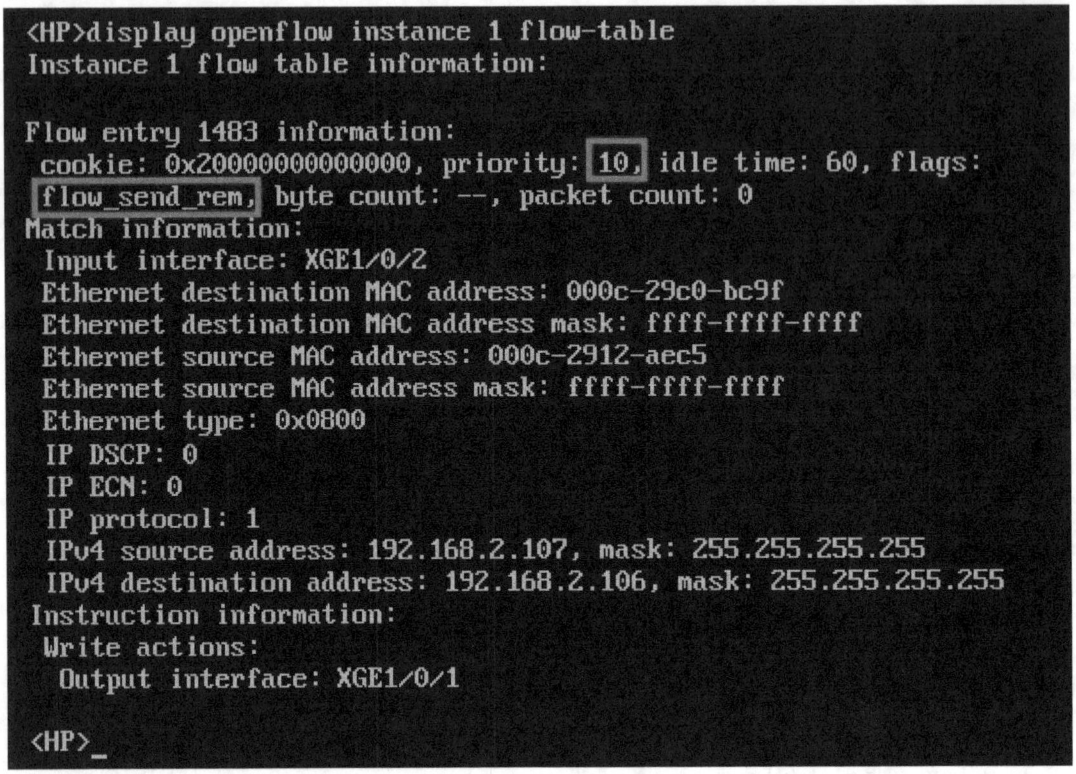

```
<HP>display openflow instance 1 flow-table
Instance 1 flow table information:


Flow entry 1483 information:
 cookie: 0x20000000000000, priority: 10, idle time: 60, flags:
 flow_send_rem, byte count: --, packet count: 0
Match information:
 Input interface: XGE1/0/2
 Ethernet destination MAC address: 000c-29c0-bc9f
 Ethernet destination MAC address mask: ffff-ffff-ffff
 Ethernet source MAC address: 000c-2912-aec5
 Ethernet source MAC address mask: ffff-ffff-ffff
 Ethernet type: 0x0800
 IP DSCP: 0
 IP ECN: 0
 IP protocol: 1
 IPv4 source address: 192.168.2.107, mask: 255.255.255.255
 IPv4 destination address: 192.168.2.106, mask: 255.255.255.255
Instruction information:
 Write actions:
  Output interface: XGE1/0/1

<HP>_
```

Figure 4.7: HP Switch Flow State With Ryu Controller Attached

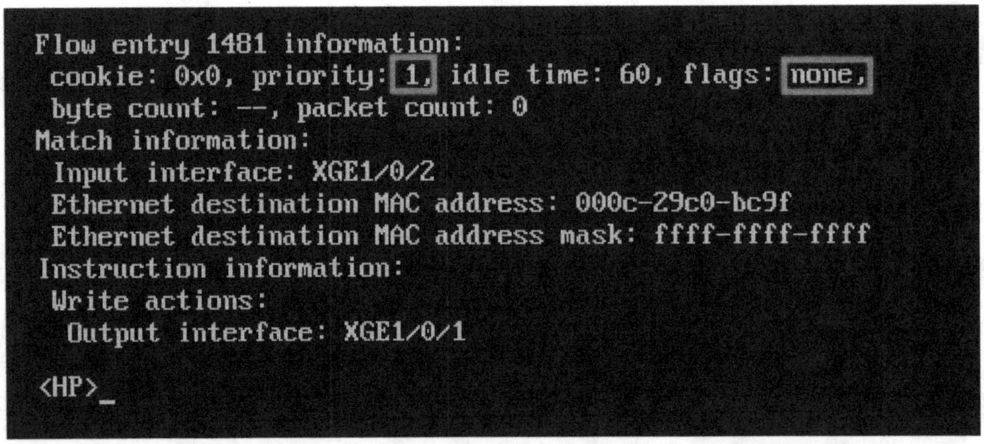

```
Flow entry 1481 information:
 cookie: 0x0, priority: 1, idle time: 60, flags: none,
 byte count: --, packet count: 0
Match information:
 Input interface: XGE1/0/2
 Ethernet destination MAC address: 000c-29c0-bc9f
 Ethernet destination MAC address mask: ffff-ffff-ffff
Instruction information:
 Write actions:
  Output interface: XGE1/0/1

<HP>_
```

Figure 4.8: HP Switch Flow State With Iris Controller Attached

Determining the inactivity timeout using the method in Experiment 2 is unreliable as the value obtained is dependent on the network switch's implementation of the Openflow protocol. The inactivity value obtained is the true length of time that the flow remains active within the HP switch, rather than the desired default time set by the SDN controller. While the true length of time is information that is potentially helpful in the discovery of DoS vulnerabilities, such research is beyond the scope of this thesis and is suggested for future work.

4.3 Experiment 3 Results and Analysis

The purpose of Experiment 3 is to determine whether a host can successfully determine the controller's hard timeout of the SDN environment. Section 4.3.1 presents the data collected using the MiniNet simulation, while Section 4.3.2 presents the data collected using the hardware switch and controller.

4.3.1 Simulated Data.

Table 4.15 shows a subset of the collected data. The first column specifies the controller used in the SDN environment. The second column is the hard timeout value in seconds that was programmed into the controller. The third column displays the response variable: the predicted hard timeout value in seconds obtained by the procedure outlined in Chapter 3. The fourth column shows the difference between the predicted hard timeout and the actual hard timeout values. The difference in values shows how accurate the prediction method was in determining the actual hard timeout value, where a smaller difference represents a better estimate. The fifth column is the elapsed runtime in seconds for obtaining the predicted hard timeout value.

Table 4.16 shows a summary of the collected data for Experiment 3. The controller column represents the specific controller in use while the data was collected. Under column controller, the term "combined" represents all the data combined, while "combined (no

Table 4.15: Experiment 3 Sample Data

Controller	Actual Timeout (s)	Predicted Timeout (s)	Δ(s)	Runtime (s)
POX	10	10.80451	0.80451	9.723884
POX	180	180.84806	0.84806	179.762805
Beacon	10	10.77271	0.77271	9.683617
Beacon	180	180.86693	0.86693	179.782083
Maestro	10	11.13056	1.13056	10.068902
Maestro	180	180.85915	0.85915	179.818652

outliers)" includes all data minus the outliers shown in the histogram shown in Figure 4.9. The Δ_{min} column shows the minimum difference observed between the predicted hard timeout and actual hard timeout values, while the Δ_{max} column shows the maximum difference observed between the predicted hard timeout and actual hard timeout values. The $\overline{\Delta}$ column shows the mean difference observed between the predicted timeout and actual timeout values, and the s_Δ column shows the sample standard deviation.

Table 4.16: Experiment 3 Summary Data

Controller	$\Delta_{min}(s)$	$\Delta_{max}(s)$	$\overline{\Delta}(s)$	s_Δ
POX	0.520713	155.47555	8.5625996	34.58039
Beacon	0.198166	104.06397	5.977903	23.08955
Maestro	0.256199	335.683227	34.726516	83.9809
Combined	0.198166	335.683227	16.97597	58.93733
Combined (no outliers)	0.198166	1.412594	0.8285399	0.3248459

Figure 4.9 displays a histogram showing the distribution of values for the timeout delta. Given 7 outliers exist in Figure 4.9, Figure 4.10 is generated to view the distribution

90

of hard timeout delta values among the 73 values between zero and ten. The affect of the outliers as well as attribution to the cause of the outliers is discussed in Section 4.3.1.1.

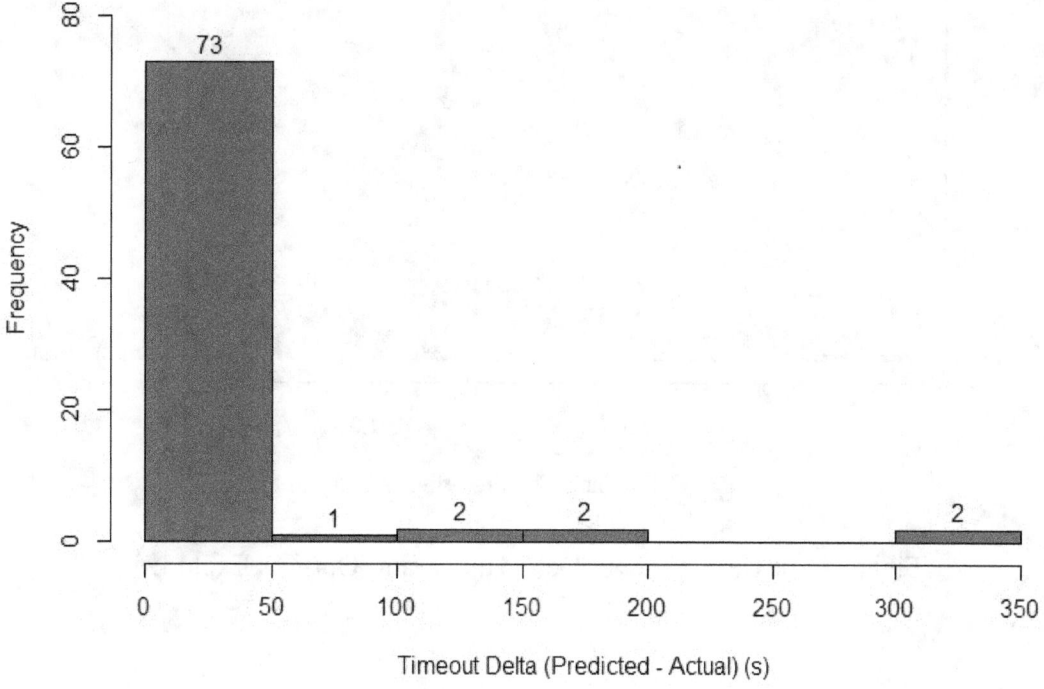

Figure 4.9: Hard Timeout Delta Histogram (Outliers Included)

4.3.1.1 Analysis.

Determining a relationship between two samples involves calculating Pearson's r correlation coefficient, which is a value between -1 and 1 inclusive and describes the degree of linear association between two variables [50]. Values close to zero indicate no correlation, and thus no observable relationship between the two variables, while values near positive or negative one show a stronger correlation between both variables, and thus a strong observable relationship between both variables. The equation for Pearson's r correlation coefficient is the same equation used in the analysis for Experiment 2 and is shown under Section 4.2.1.1 in (4.2).

91

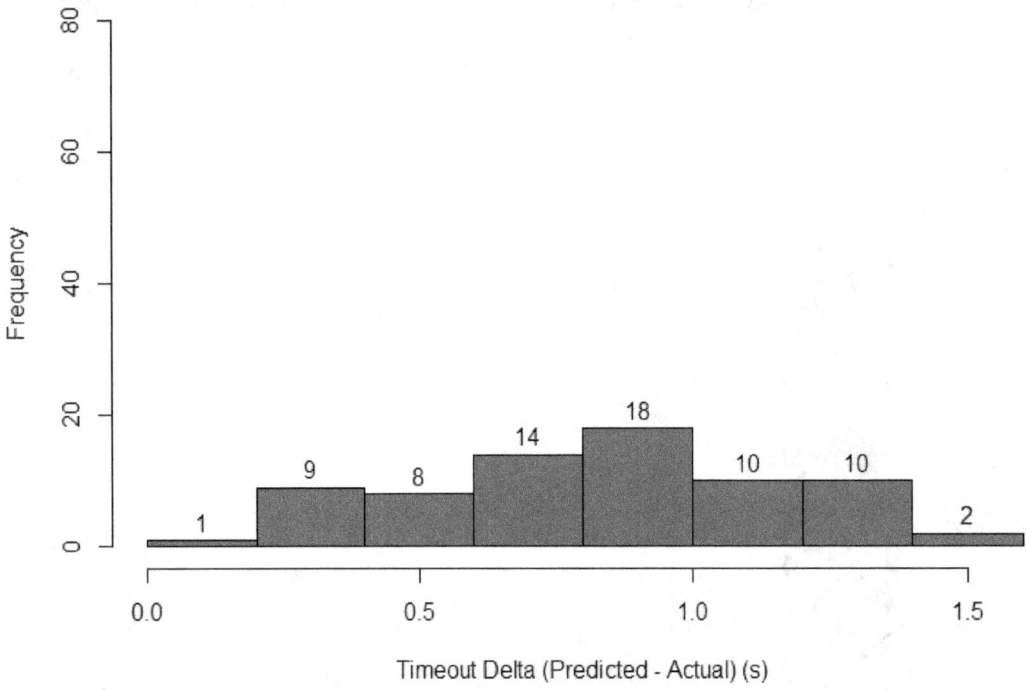

Figure 4.10: Hard Timeout Delta Histogram (Outliers Excluded)

Applying (4.2) to the data in Experiment 3, X represents the sample of actual hard timeout values that were programmed into the controller, while Y represents the response variable sample: the predicted hard timeout values. The value n represents the sample size of both X and Y combined, while s represents the sample standard deviation.

Pearson's r correlation coefficient for both sample sets, with outliers included, is 0.8872715, which indicates a strong positive correlation between the predicted hard timeout value and the actual hard timeout value. The existence of a strong correlation, along with a mean delta of 16.97597 seconds, indicates that the method of predicting the actual hard timeout value has a precision of less than 17 seconds. The standard deviation of both sample sets is 58.93733, which indicates a lack of reliability in determining the hard timeout with a precision of less than 1 second.

Outliers heavily affect the mean, standard deviation, and correlation coefficient. With the seven outliers omitted, the coefficient increases to 0.9993395, with a mean delta of 0.8285399 and a standard deviation of 0.3248459, indicating strong precision in predicting the actual hard timeout value for a controller. A 95% confidence interval of the mean delta value with outliers omitted is between 0.75220 and 0.90488.

Outliers can be attributed to network latency spikes. Considering that the method for determining the hard timeout relies on taking an initial sample of RTT values for an ICMP packet, then detecting a spike in a new RTT value, a spike due to normal networking lag rather than the installation of a new flow would cause the method to generate a false predicted hard timeout value. Depending on when the spike occurred, the resulting delta may be anywhere between zero and the actual hard timeout value. Considering that the method for determining the hard timeout involves routinely transmitting an ICMP echo packet, and comparing its RTT with the baseline RTT, the chances of an RTT spike due to normal network activity (rather than a flow installation event) increases with an increased hard timeout value. Additionally, given that the hard timeout value is usually greater than the inactivity timeout, it is more likely for greater outliers to exist with hard timeout values.

4.3.1.2   Other Observations.

An interesting observation to note includes the runtime required to obtain the hard timeout values. In order to determine that a test value is greater than the true hard timeout value, the elapsed time must be greater than or equal to the test value. Because of the elapsed time requirement, testing for 10 seconds requires at least 10 seconds for the flow to expire from a hard timeout. Once the hard timeout event occurs, no further testing is required, making the elapsed time linearly related to the actual hard timeout value.

4.3.2   Hardware Data.

Table 4.17 shows a subset of the collected data. The first column specifies the controller used in the SDN environment. The second column is the hard timeout value in seconds that was programmed into the controller. The third column displays the response variable: the predicted hard timeout value in seconds obtained by the procedure outlined in Chapter 3. The fourth column shows the absolute value of the difference between the predicted hard timeout and the actual hard timeout values. The difference in values shows how accurate the prediction method was in determining the actual hard timeout value, where a smaller difference represents a better estimate.

Table 4.17: Experiment 3 Sample Hardware Data

Controller	Actual Timeout (s)	Predicted Timeout (s)	$|\Delta|$(s)
Ryu	10	11.28655	1.28655
Ryu	180	180.84806	0.84806
Iris	10	10.77637	0.77637
Iris	180	180.6244	0.6244

Table 4.18 shows a summary of the collected data for Experiment 3. The controller column represents the specific controller in use while the data was collected. Under column controller, the term "combined" represents all the data combined, while "combined (no outliers)" includes all data minus the one outlier shown in the histogram described by Figure 4.11.

The Δ_{min} column shows the minimum difference observed between the predicted hard timeout and actual hard timeout values, while the Δ_{max} column shows the maximum difference observed between the predicted hard timeout and actual hard timeout values. The $\overline{\Delta}$ column shows the mean difference between the predicted hard timeout values and the actual hard timeout values. The s_Δ column shows the standard deviation of the differences.

Figure 4.11 displays a histogram showing the distribution of values for the hard timeout delta. Given 1 outlier exists in Figure 4.11, Figure 4.10 is generated to view

Table 4.18: Experiment 3 Summary Hardware Data

Controller	$\Delta_{min}(s)$	$\Delta_{max}(s)$	$\overline{\Delta}(s)$	s_Δ
Ryu	0.0059646	286.0684	14.5653	63.90537
Iris	0.0006436	0.6344986	0.2513435	0.1505713
Combined	0.0006436	286.0684	7.408323	45.19001
Combined (no outliers)	0.0006436	0.6344986	0.2631929	0.1499256

the distribution of hard timeout delta values among the 39 values between zero and fifty. The affect of the outlier as well as attribution to the cause of the outlier is discussed in Section 4.3.2.1.

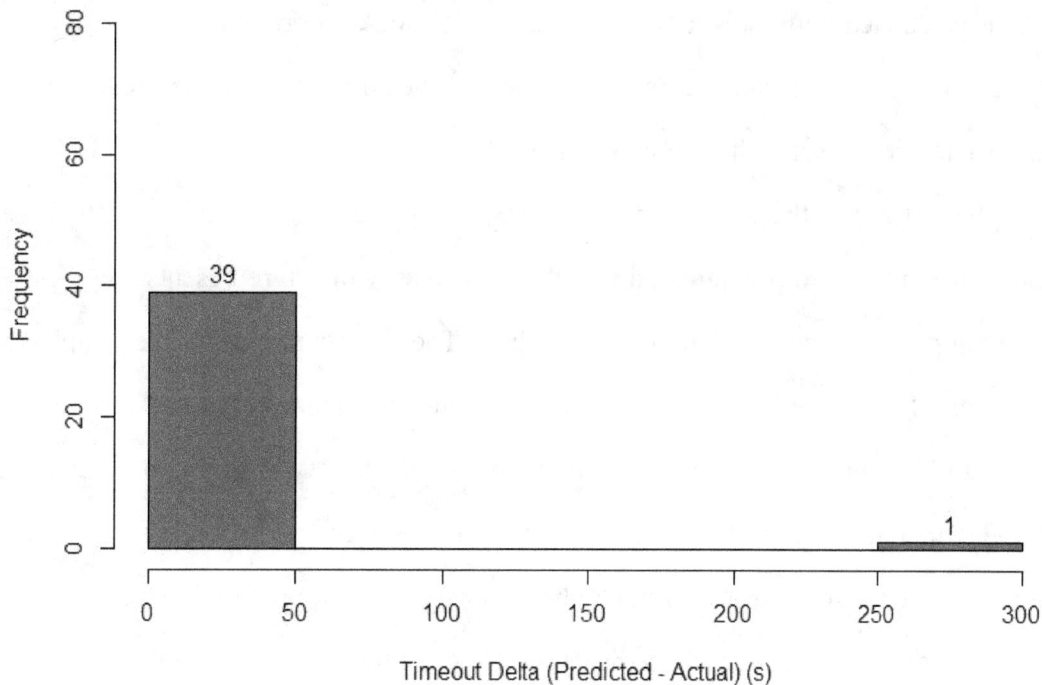

Figure 4.11: Hard Timeout Deltas Using The Hardware HP Switch (Outlier Included)

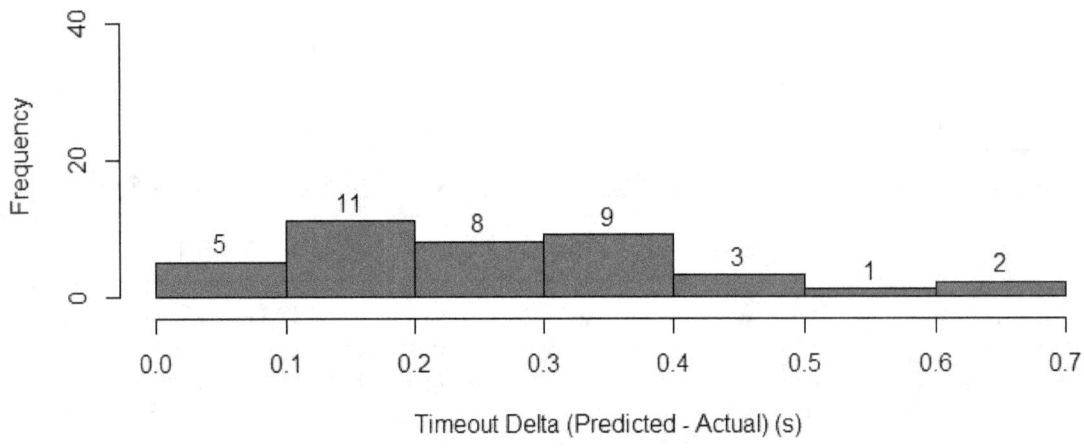

Figure 4.12: Hard Timeout Deltas Using The Hardware HP Switch (Outliers Excluded)

4.3.2.1 *Analysis.*

As is calculated with the simulation data, the Pearson's r correlation coefficient from (4.2) is again used to describe the degree of linear association between the predicted hard timeout value and the actual hard timeout value.

Applying (4.2) to the data in Experiment 3, X represents the sample of actual hard timeout values that were programmed into the controller, while Y represents the response variable sample: the predicted hard timeout values. The value n represents the sample size of both X and Y combined, while s represents the sample standard deviation.

Pearson's r correlation coefficient for both sample sets, with outliers included, is 0.934193, which indicates a strong positive correlation between the predicted hard timeout value and the actual hard timeout value. The existence of a strong correlation, along with a mean delta of 7.408323 seconds, indicates that the method of predicting the actual hard timeout value has a precision of less than 8 seconds. The standard deviation of both sample sets is 45.19001, which indicates a lack of reliability in determining the hard timeout with a precision of less than 1 second.

96

Outliers heavily affect the mean, standard deviation, and correlation coefficient. With the single outlier omitted, the coefficient increases to 0.999997, with a mean delta of 0.2631929 and a standard deviation of 0.1499256, indicating strong precision (i.e., a precision of near 1 second) in predicting the actual hard timeout value for a controller. A 95% confidence interval of the mean delta value with outliers omitted is between 0.21459 and 0.31179.

Similar to the simulated data, outliers can be attributed to network latency spikes. Considering that the method for determining the hard timeout relies on taking an initial sample of RTT values for an ICMP packet, then detecting a spike in a new RTT value, a spike due to normal networking lag rather than the installation of a new flow would cause the method to generate a false predicted hard timeout value. Depending on when the spike occurred, the resulting delta may be anywhere between zero and the actual hard timeout value. Considering that the method for determining the hard timeout involves routinely transmitting an ICMP echo packet, and comparing its RTT with the baseline RTT, the chances of an RTT spike due to normal network activity (rather than a flow installation event) increase with an increased actual hard timeout value. Additionally, given that the hard timeout value is usually greater than the inactivity timeout, it is more likely for greater outliers to exist with hard timeout values.

4.3.2.2 Other Observations.

An interesting observation to note is that every predicted timeout value obtained is greater than the actual timeout. The consistency in which the predicted timeout is greater than the actual timeout value indicates that the method can be linearly optimized by simply subtracting a constant amount each time the predicted value is obtained. The constant amount selected is the mean of the delta values obtained (i.e., 0.9630554). While this is not the optimum value for all cases, it provides a better prediction of the actual inactivity timeout value. Applying this subtraction to all data for both Ryu and Iris controllers, along with the removal of the single outlier, reduces the mean delta to 0.2631929 which is well within our desired precision of one second.

Achieving the desired precision of less than one second in determining the hard timeout value indicates that the method for obtaining the hard timeout value is reliable. However, the previous experiment (i.e., Experiment 2) raises concerns regarding the influence of the network switch, (i.e., the introduction of the network switch as a variable). While the hard timeout value is strictly enforced with the HP switch, the results of Experiment 3 are limited to interpolation within a netork containing the specific HP 5900 series Openflow switch. A separate network may introduce a different Openflow network switch that does not enforce the hardware timeout value as precisely as the HP switch, and consequently may provide different results.

4.4 Experiment 4 Results and Analysis

The purpose of Experiment 4 is to determine which SDN controller is creating and managing flows for the network switch. Section 4.4.1 describes why the data for the network designed in Chapter 3 are not observable.

4.4.1 Result.

Within MiniNet, the communication between each end host is accomplished through virtual ports connected to OpenVSwitch. By default, the end host does not see the the remote SDN controller's network. Without a communication link between a host on the network and the SDN controller, an assumption is violated, and the methods of crafting and transmitting custom OpenFlow packets to an unknown controller become impossible. The features available for extracting are limited to features observable from the host's perspective (i.e., the data collected from Experiments 1-3). For this thesis, these features include whether the network is an SDN or traditional environment, the flow inactivity timeout value, and the flow hard timeout value. The use of only three features is insufficient to uniquely identify an SDN controller considering the likelihood for the three features to be identical among any two controllers is high.

The result of the simulated network extends to the hardware network. The HP switch uses a Virtual Local Area Network (VLAN) to segregate control traffic (i.e., traffic between the controller and the OpenFlow switch) from other network traffic. The isolation of control traffic prohibits any host attached to the HP switch from communicating directly with the controller, and consequently limits the number of features a host can collect from the SDN controller.

4.4.2 Observations.

While a host is unable to communicate directly with the controller attached to the SDN environment due to network segregation, information can be obtained from a separate controller found within the same network. Reasons for an SDN controller existing on the network include a network administrator testing an SDN controller that is not yet deployed, or an older version of an SDN controller that was moved from an isolated network to another network. In either case, gleaning information about visible SDN controllers gives a host insight into the potential SDN controller managing the current network.

The following data is collected from a modification to the original network environment described in Chapter 3. The new network environment is depicted in Figure 4.13, and is similar to the original network environment with the exception that the unknown "discovered controller" is found within the same VLAN and is thus visible from the perspective of a network host. The controller managing the SDN environment in Figure 4.13 exists within a separate VLAN as that of the hosts, while the newly discovered controller exists within the same VLAN as the host. The discovered controller is an SDN controller running on top of an Ubuntu 12.04 Linux Virtual Machine.

Ensuring that a connection exists to a controller fulfills the assumption of a communication link between the host and controller. Table 4.19 explains each feature that

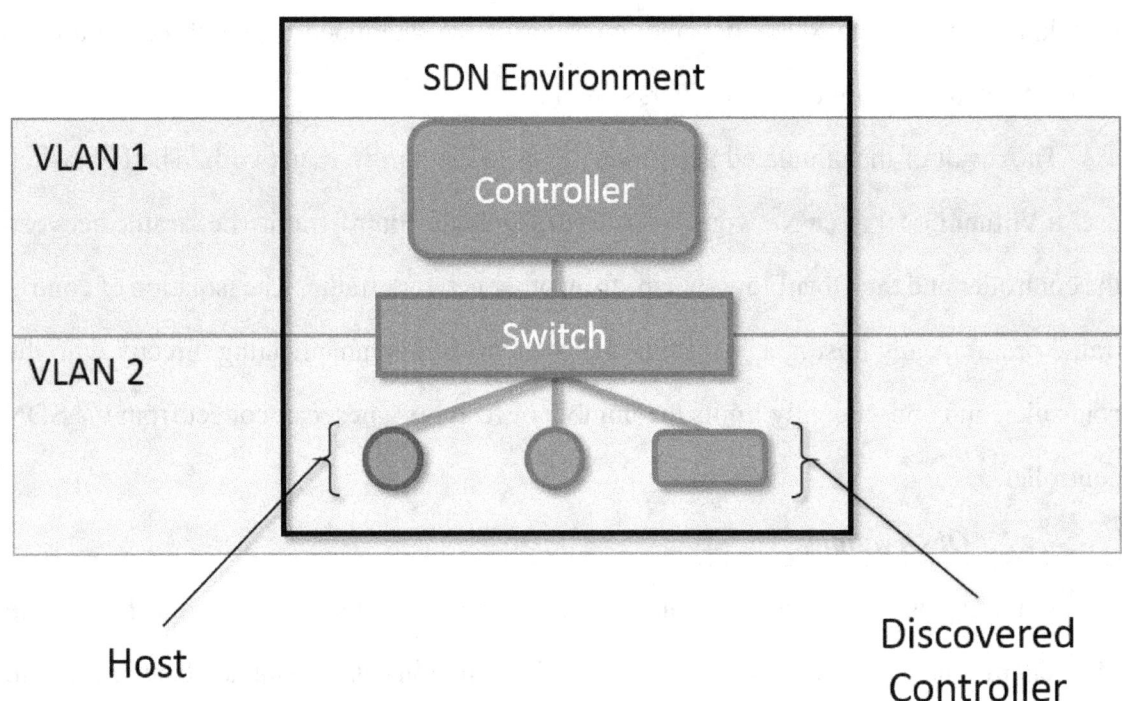

Figure 4.13: Modified SDN Simple Network

is captured on the newly discovered SDN controller. The left-most column represents the identification number of the feature, and corresponds to the same identification numbers 1 through 7 in Table 4.20.

Table 4.20 shows the features of the discovered controller collected by the host in the network. The actual column represents the actual controller that is running on the Ubuntu virtual machine. The predicted column shows the response variable: the predicted controller operating on the Ubuntu virtual machine. Columns titled 1 through 7 correspond to each feature number listed in Table 4.19. Columns 6 and 7 show a list of numbers that correspond to the OpenFlow packet type listed in Table 3.16. For columns 6 and 7, a list within parenthesis means that both OpenFlow packets were sent within the same Ethernet frame, while numbers separated by a comma represent an ordered event. For example,

100

Table 4.19: List of Features

ID	Feature
1	Default Flow Inactivity-Timeout Period
2	Default Flow Hard-Timeout Period
3	Does the SDN controller initiate an OF Hello packet?
4	Does the SDN controller initiate an Echo Request?
5	Does the SDN controller initiate a Feature Request?
6	The set of packets sent by the SDN controller just after the Hello exchange
7	The set of packets sent by the SDN controller just after the Feature Response

the very last row and last column (i.e., 9,(14,18)), represents an OpenFlow packet type 9 arriving, followed by a single Ethernet frame containing both OpenFlow packet types 14 and 18, respectively. OpenFlow packet type 9 represents a "Set Configuration" packet, and OpenFlow packet types 14 and 18 represent a "Flow Modification" and "Barrier Request" packet type, respectively (as specified in Table 3.16).

Table 4.20 shows that every time the controller was randomly set, the method of predicting the controller was accurate. In all cases except three, every feature of the unknown controller was correctly matched with the feature of the predicted controller. In Table 4.20, row 6 (i.e., the first Maestro controller), an incorrect value of 63.996s for the flow hard timeout was found. The value of 63.996s for a default flow hard timeout value was not found in the table for any controller. Because a perfect match does not exist, the set of features of each controller in the table was compared with each other. The controller that has the highest number of matching features was selected. The same best-guess action was taken for rows 15 and 16 in Table 4.20, as either the flow inactivity timeout, or flow hard timeout default values were not exactly paired with the known default value.

Table 4.20: SDN Controller Feature Table

Actual	Predicted	1	2	3	4	5	6	7
NOX	NOX	5.625	>181	False	False	False	(5,9)	None
Nodeflow	Nodeflow	5.625	>181	False	False	False	0,5	None
NOX	NOX	5.375	>181	False	False	False	(5,9)	None
OpenDaylight	OpenDaylight	>120	>181	True	False	False	5,5,5	9,7,14
Nodeflow	Nodeflow	5.625	>181	False	False	False	0,5	None
Maestro	Maestro	30.625	63.996	True	False	False	5	(13,13,13,13)
POX	POX	10.375	30.784	True	False	False	5	9,(14,18)
Beacon	Beacon	5.375	>181	True	True	False	5,2,2	16,2,2
Nodeflow	Nodeflow	5.375	>181	False	False	False	0,5	None
Nodeflow	Nodeflow	5.375	>181	False	False	False	0,5	None
POX	POX	10.375	30.741	True	False	False	5	9,(14,18)
Maestro	Maestro	30.375	180.343	True	False	False	5	(13,13,13,13)
Beacon	Beacon	5.625	>181	True	True	False	5,2,2	16,2,2
Floodlight	Floodlight	5.875	>181	True	False	False	5	(9,18,7)
Floodlight	Floodlight	6.125	>181	True	False	False	5	(9,18,7)
Maestro	Maestro	5.625	180.347	True	False	False	5	(13,13,13,13)
POX	POX	10.375	30.812	True	False	False	5	9,(14,18)

The fact that all cases of controllers are accurately predicted shows that this method of identifying an unknown controller from a table of known controllers is accurate provided the controller exists within the table, and provided that the controller features are extractable. The requirement that the controller be in the table enforces the need for a growing table of many versions of each type of controller.

V. Conclusions

This chapter provides a summary for the results of each experiment, and explains the relation each result has to the goal of this research. Section 5.1 summarizes each conclusion drawn from each experiment. Next, Section 5.2 explains the significance this research has on the body of research for software defined networking. Finally, Section 5.3 presents suggested future work to continue this research.

5.1 Research Conclusions

5.1.1 Goal #1: Construct a table of features.

The first goal of this research is to construct a set of features extensive enough to uniquely identify each known SDN controller, and demonstrate that the table of features reliably identifies each SDN controller. Table 4.20 from Section 4.4.2 shows that the set of features selected for each SDN controller is extensive enough to uniquely identify each SDN controller. The method of creating a table of features requires creating a new feature when more than one SDN controller contains an exact match of features. Creating a new feature ensures that each SDN controller remains distinct within the table. The results of Experiment 4 with a communication link between the host and the SDN controller demonstrate that the table of features correctly differentiated between known controllers.

5.1.2 Goal #2: Verifying feature extraction.

The second goal includes ensuring that each feature from goal 1 is obtainable by a host connected to the SDN environment. The second goal requires that the environment be verified before proceeding to collect features. The results of Experiment 1 demonstrate that it is possible to distinguish between a traditional network environment and an SDN environment. The results of Experiment 1 apply to both the simulated MiniNet environment as well as the emulated environment using the HP switch. When simulated using MiniNet,

103

the results of Experiment 2 demonstrate that one feature, the flow's inactivity timeout value, is obtainable by an end host of the SDN environment. The results of Experiment 2 applied to the emulated environment however show that the response inactivity timeout depends on not only the controller, but also the switch implementing the OpenFlow protocol. Experiment 2 showed that the inactivity timeout is not reliably obtainable considering the value depends on the implementation of the network switch.

When simulated in MiniNet, the results of Experiment 3 demonstrate that another feature, the flow's hard timeout value, is obtainable by an end host of the SDN environment. The results of Experiment 3 applied to the emulated environment also support the conclusions established from the simulated data. Experiment 2 sheds light on the importance of the network switch's OpenFlow implementation, and thus shows that the results of Experiment 3, while conclusive, cannot be extrapolated to other network hardware switches. Experiment 3's results are thus confined to the specific switch used during data collection (i.e., the HP 5900 series switch). The results of both Experiment 2 and Experiment 3 fail to achieve the second goal set out in this research, as both experiments demonstrate that a host, without prior knowledge of the network switch's implementation of OpenFlow, fail to reliably determine the flow timeout values set by the controller.

5.2 Research Significance

This research provides a reliable method for verifying the existence of an SDN environment. Additionally, this research presents a method for fingerprinting various SDN controllers. While the features collected from Experiments 2 and 3 are not reliably collectable from an end host connected to the SDN environment, the framework is laid out for future features to be discovered and added to the list of observable features. Discovering unique features observable by an end host brings this research closer to uniquely identifying the SDN controller. Once an SDN controller is identified, known vulnerabilities may exist for the controller and can be selected for use in exploiting the SDN network. The method of fingerprinting SDN controllers, followed by targeted exploits, parallels current network attack methodologies: first network reconnaissance identifies a target, then a vulnerability is selected from a database pertaining to the selected target, and finally a network attack ensues. Future research includes current research dedicated to these network attack methodologies, but applied specifically to SDN environments.

5.3 Future Work

Extensions to this research include discovery of new observable features. With a greater availability of new features, the feature table will more uniquely identify each controller based on the entire list of features. Additionally, methods of fingerprinting OpenFlow enabled network switches allows an increased ability in fingerprinting the OpenFlow network. Once identified, switches that behave in an expected manner according to their implementation of OpenFlow will then be removed as a variable hindering the fingerprinting of the SDN controller.

As with every network attack discovery, the idea of mitigation grows apparent. Preventing disclosure of the network environment through the use of aggregate flows or intentionally induced latency provides another avenue of research extending from this topic.

Another extension to this research includes applying the results of Experiment 2 with the creation of a targeted DoS attack. Considering Experiment 2 effectively determines how long an active flow resides within the network switch, it may be possible to create a flood of traffic designed to fill the flow table at exactly the rate of traffic necessary to disable network services. The fact that the flooding of packets is only at the level necessary to deny service, such a level may exist below the level of detection by an IDS.

105

Bibliography

[1] M. E. Dempsey, "Quadrennial defense review," 2014. Available at http://www.defense.gov/pubs/2014_Quadrennial_Defense_Review.pdf (Retrieved on 18 December 2014).

[2] S. Corporation, "Internet security threat report 2013 trends," *Internet Security Threat Report volume 19*, Apr. 2014. Available at http://www.symantec.com/content/en/us/enterprise/other_resources/b-istr_main_report_v19_21291018.en-us.pdf (Retrieved on 18 December 2014).

[3] Anonymous, "Corporate cyber-ssecurity horror movie: Hackers shine a harsh spotlight on sony," *The Economist*, Dec. 2014. Available at http://abcnews.go.com/Technology/wireStory/sony-attack-hackers-employees-27653706 (Retrieved on 18 December 2014).

[4] B. Solomon, "Apple admits celebrity photos were stolen in targeted hack," *Forbes*, Sept. 2014. Available at http://www.forbes.com/sites/briansolomon/2014/09/02/apple-admits-celebrity-photos-were-stolen-in-targeted-hack/ (Retrieved on 18 December 2014).

[5] R. Abrams, "Target puts data breach costs at $148 million, and forecasts profit drop," *The New York Times*, Aug. 2014. Available at http://www.nytimes.com/2014/08/06/business/target-puts-data-breach-costs-at-148-million.html/ (Retrieved on 18 December 2014).

[6] "Critical security controls for effective cyber defense," *SANS Critical Security Controls*. Available at https://www.sans.org/critical-security-controls/ (Retrieved on 18 December 2014).

[7] L. Bilge and T. Dumitras, "Before we knew it: an empirical study of zero-day attacks in the real world," in *Proceedings of the 2012 ACM conference on Computer and communications security*, pp. 833–844, ACM, 2012.

[8] H. Kim and N. Feamster, "Improving network management with software defined networking," *Communications Magazine, IEEE*, vol. 51, no. 2, pp. 114–119, 2013.

[9] T. D. Nadeau and K. Gray, *SDN: Software Defined Networks.* " O'Reilly Media, Inc.", 2013.

[10] V. Shukla, *Introduction to Software Defined Networking - Openflow & VxLAN.* "CreateSpace Independent Publishing Platform", 2013.

[11] N. McKeown, T. Anderson, H. Balakrishnan, G. Parulkar, L. Peterson, J. Rexford, S. Shenker, and J. Turner, "Openflow: enabling innovation in campus networks," *ACM SIGCOMM Computer Communication Review*, vol. 38, no. 2, pp. 69–74, 2008.

[12] "Openflow switch specification," Mar. 2014. Available at https://www.opennetworking.org/sdn-resources/onf-specifications/openflow (Retrieved on 1 October 2014).

[13] A. R. Curtis, J. C. Mogul, J. Tourrilhes, P. Yalagandula, P. Sharma, and S. Banerjee, "Devoflow: scaling flow management for high-performance networks," in *ACM SIGCOMM Computer Communication Review*, vol. 41, pp. 254–265, ACM, 2011.

[14] N. Gude, T. Koponen, J. Pettit, B. Pfaff, M. Casado, N. McKeown, and S. Shenker, "Nox: towards an operating system for networks," *ACM SIGCOMM Computer Communication Review*, vol. 38, no. 3, pp. 105–110, 2008.

[15] D. Erickson, "The beacon openflow controller," in *Proceedings of the Second ACM SIGCOMM Workshop on Hot Topics in Software Defined Networking*, pp. 13–18, ACM, 2013.

[16] Z. Cai, A. L. Cox, and T. E. N. Maestro, "A system for scalable openflow control," tech. rep., Technical Report TR10-08, Rice University, 2010. Available at http://www.cs.rice.edu/ eugeneng/papers/TR10-11.pdf (Retrieved on 1 October 2014).

[17] A. Tootoonchian, S. Gorbunov, Y. Ganjali, M. Casado, and R. Sherwood, "On controller performance in software-defined networks," in *Proceedings of the 2nd USENIX Conference on Hot Topics in Management of Internet, Cloud, and Enterprise Networks and Services*, Hot-ICE'12, (Berkeley, CA, USA), pp. 10–10, USENIX Association, 2012. Available at http://dl.acm.org/citation.cfm?id=2228283.2228297 (Retrieved on 1 October 2014).

[18] "Project floodlight," 2014. Available at http://www.projectfloodlight.org/floodlight/ (Retrieved on 21 December 2014).

[19] "Big switch networks." Available at: http://www.bigswitch.com/ (Retrieved on 13 October 2014).

[20] "Loxigen." Available at: http://github.com/floodlight/loxigen/wiki/OpenFlowJ-Loxi (Retrieved on 13 October 2014).

[21] B. Lee, S. H. Park, J. Shin, and S. Yang, "Iris: The openflow-based recursive sdn controller," in *Advanced Communication Technology (ICACT), 2014 16th International Conference on*, pp. 1227–1231, IEEE, 2014.

[22] A. Tootoonchian and Y. Ganjali, "Hyperflow: A distributed control plane for openflow," in *Proceedings of the 2010 Internet Network Management Conference on Research on Enterprise Networking*, INM/WREN'10, (Berkeley, CA, USA), pp. 3–3, USENIX Association, 2010. Available at http://dl.acm.org/citation.cfm?id=1863133.1863136 (Retrieved on 1 October 2014).

[23] A. Shalimov, D. Zuikov, D. Zimarina, V. Pashkov, and R. Smeliansky, "Advanced study of sdn/openflow controllers," in *Proceedings of the 9th Central & Eastern European Software Engineering Conference in Russia*, p. 1, ACM, 2013.

[24] A. Greenhalgh, F. Huici, M. Hoerdt, P. Papadimitriou, M. Handley, and L. Mathy, "Flow processing and the rise of commodity network hardware," *ACM SIGCOMM Computer Communication Review*, vol. 39, no. 2, pp. 20–26, 2009.

[25] B. Lantz, B. Heller, and N. McKeown, "A network in a laptop: rapid prototyping for software-defined networks," in *Proceedings of the 9th ACM SIGCOMM Workshop on Hot Topics in Networks*, p. 19, ACM, 2010.

[26] M. Gupta, J. Sommers, and P. Barford, "Fast, accurate simulation for sdn prototyping," in *Proceedings of the Second ACM SIGCOMM Workshop on Hot Topics in Software Defined Networking*, pp. 31–36, ACM, 2013.

[27] S. McCanne, S. Floyd, K. Fall, K. Varadhan, *et al.*, "Network simulator ns-2," 1997.

[28] T. R. Henderson, M. Lacage, G. F. Riley, C. Dowell, and J. Kopena, "Network simulations with the ns-3 simulator," *SIGCOMM demonstration*, 2008.

[29] S. Jain, A. Kumar, S. Mandal, J. Ong, L. Poutievski, A. Singh, S. Venkata, J. Wanderer, J. Zhou, M. Zhu, *et al.*, "B4: Experience with a globally-deployed software defined wan," in *Proceedings of the ACM SIGCOMM 2013 conference on SIGCOMM*, pp. 3–14, ACM, 2013.

[30] U. Hoelzle, "Openflow @ google." Available at http://www.opennetsummit.org/archives/apr12/hoelzle-tue-openflow.pdf (Retrieved on 13 October 2014).

[31] M. Palmer, "Sdncentral exclusive: Sdn market size expected to reach 35b by 2018," 2013. Available at https://www.sdncentral.com/market/sdn-market-sizing/2013/04/ (Retrieved on 13 October 2014).

[32] R. Klöti, "Openflow: A security analysis," *Proc. Workshop on Secure Network Protocols (NPSec). IEEE*, 2013. Available at http://www.csg.ethz.ch/people/vkotroni/openflow_sec.

[33] S. Hernan, S. Lambert, T. Ostwald, and A. Shostack, "Threat modeling-uncover security design flaws using the stride approach," *MSDN Magazine-Louisville*, pp. 68–75, 2006.

[34] V. Saini, Q. Duan, and V. Paruchuri, "Threat modeling using attack trees," *Journal of Computing Sciences in Colleges*, vol. 23, no. 4, pp. 124–131, 2008.

[35] S. Scott-Hayward, G. O'Callaghan, and S. Sezer, "Sdn security: A survey," in *Future Networks and Services (SDN4FNS), 2013 IEEE SDN for*, pp. 1–7, IEEE, 2013.

[36] S. Shin and G. Gu, "Attacking software-defined networks: A first feasibility study," in *Proceedings of the Second ACM SIGCOMM Workshop on Hot Topics in Software Defined Networking*, pp. 165–166, ACM, 2013.

[37] J. M. Dover, "A switch table vulnerability in the open floodlight sdn controller," Available at http://dovernetworks.com/wp-content/uploads/2014/03/OpenFloodlight-03052014.pdf.

[38] J. M. Dover, "A denial of service attack against the open floodlight sdn controller," Available at http://dovernetworks.com/wp-content/uploads/2013/12/OpenFloodlight-12302013.pdf.

[39] D. Kreutz, F. Ramos, and P. Verissimo, "Towards secure and dependable software-defined networks," in *Proceedings of the Second ACM SIGCOMM Workshop on Hot Topics in Software Defined Networking*, pp. 55–60, ACM, 2013.

[40] S. A. Mehdi, J. Khalid, and S. A. Khayam, "Revisiting traffic anomaly detection using software defined networking," in *Recent Advances in Intrusion Detection*, pp. 161–180, Springer, 2011.

[41] R. Braga, E. Mota, and A. Passito, "Lightweight ddos flooding attack detection using nox/openflow," in *Local Computer Networks (LCN), 2010 IEEE 35th Conference on*, pp. 408–415, IEEE, 2010.

[42] T. Xing, D. Huang, L. Xu, C.-J. Chung, and P. Khatkar, "Snortflow: A openflow-based intrusion prevention system in cloud environment," in *Research and Educational Experiment Workshop (GREE), 2013 Second GENI*, pp. 89–92, IEEE, 2013.

[43] S. Shin, P. A. Porras, V. Yegneswaran, M. W. Fong, G. Gu, and M. Tyson, "Fresco: Modular composable security services for software-defined networks," in *NDSS*, The Internet Society, 2013. Available at http://internetsociety.org/doc/fresco-modular-composable-security-services-software-defined-networks (Retrieved on 1 October 2014).

[44] J. H. Jafarian, E. Al-Shaer, and Q. Duan, "Openflow random host mutation: Transparent moving target defense using software defined networking," in *Proceedings of the First Workshop on Hot Topics in Software Defined Networks*, pp. 127–132, ACM, 2012.

[45] A. Zarek, Y. Ganjali, and D. Lie, "Openflow timeouts demystified," *Univ. of Toronto, Toronto, Ontario, Canada*, 2012.

[46] A. Vishnoi, R. Poddar, V. Mann, and S. Bhattacharya, "Effective switch memory management in openflow networks," in *Proceedings of the 8th ACM International Conference on Distributed Event-Based Systems*, pp. 177–188, ACM, 2014.

[47] T. Kim, K. Lee, J. Lee, S. Park, Y.-H. Kim, and B. Lee, "A dynamic timeout control algorithm in software defined networks,"

[48] Hewlett-Packard, "Quickspecs hp 5900 switch series," Dec. 2014.   Available at http://h18000.www1.hp.com/products/quickspecs/14252_na/14252_na.pdf   (Retrieved on 18 December 2014).

[49] "Ryu sdn framework," Dec. 2014.   Available at http://osrg.github.io/ryu/ (Retrieved on 18 December 2014).

[50] F. Ramsey and D. Schafer, *The statistical sleuth: a course in methods of data analysis.* Cengage Learning, 2012.

[51] Hewlett-Packard,   "Hp   5920   and   5900   switch   series: Openflow   command   reference,"   2014.   Available   at http://h20565.www2.hp.com/hpsc/doc/public/display?docId=emr_na-c04089449-1 (Retrieved on 18 December 2014).

www.ingramcontent.com/pod-product-compliance
Lightning Source LLC
Chambersburg PA
CBHW060009210526

45170CB00017B/2098